Gödel's Theorem

Gödel's Theorem

An Incomplete Guide to Its Use and Abuse

Torkel Franzén
Luleå University of Technology, Sweden

A K Peters
Wellesley, Massachusetts

Editorial, Sales, and Customer Service Office

A K Peters, Ltd.
888 Worcester Street, Suite 230
Wellesley, MA 02482
www.akpeters.com

Library of Congress Cataloging-in-Publication Data

Franzén, Torkel.
Gödel's theorem : an incomplete guide to its use and abuse / Torkel Franzén.
p. cm.
Includes bibliographical references and index.
ISBN 1-56881-238-8
1. Gödel's theorem. 2. Incompleteness theorems. I. Title.

QA9.65.F73 2005
511.3–dc22

2005045868

About the cover: Sampled on the cover are various arguments and reflections invoking Gödels theorem. The aim of the book is to allow a reader with no knowledge of formal logic to form a sober and soundly based opinion of these uses and abuses.

Printed in the United States of America
09 08 07 06 05 10 9 8 7 6 5 4 3 2

For Marcia,

still bright and beautiful, in her joy, in her despair

Contents

Preface

My excuse for presenting yet another book on Gödel's incompleteness theorem written for a general audience is that no existing book both explains the theorem from a mathematical point of view, including that of computability theory, and comments on a fairly wide selection of the many invocations of the incompleteness theorem outside mathematics.

To a considerable extent, the book reflects my experiences over the years of reading and commenting on references to the incompleteness theorem on the Internet. Quotations from named sources that as far as I know exist only in electronic form are not accompanied by any URLs, since such often become obsolete. However, using a search engine, the reader can easily locate any quoted text that is still extant somewhere on the Internet.

In quite a few cases, comments that I have encountered on the Internet in informal contexts are quoted (sometimes in slightly edited form) without attribution. They are used to represent commonly occurring ideas and arguments.

In thanking those who have helped me write this book, I must begin with the many people discussing Gödel's theorem on the Internet, whether named in the book or not, without whose contributions it is unlikely that the book would ever have appeared. I also thank Andrew Boucher, Damjan Bojadziev, Alex Blum, Jeff Dalton, Solomon Feferman, John Harrison, Jeffrey Ketland, Panu Raatikainen, and Charles Silver for helpful comments. In writing the book, I have drawn on the resources of Luleå University of Technology in several essential ways, for which I am grateful.

For any remaining instances of incompleteness or inconsistency in the book, I consider myself entirely blameless, since after all, Gödel proved that any book on the incompleteness theorem must be incomplete or inconsistent. Well, maybe not. Although the book will perhaps in part be heavy going for readers not used to mathematical proofs and definitions, I hope it will give even casual readers a basis for judging for themselves the merits of such nonmathematical appeals to the incompleteness theorem and an appreciation of some of the philosophical and mathematical perspectives opened up by the theorem.

Torkel Franzén

1

Introduction

1.1 The Incompleteness Theorem

Few theorems of pure mathematics have attracted much attention outside the field of mathematics itself. In recent years, we have seen Fermat's Last Theorem attract the interest of the general public through its much-publicized final proof by Andrew Wiles, and many nonmathematicians could probably state and illustrate the theorem of Pythagoras, about the square of the hypotenuse of a right triangle, which made it into a song performed by Danny Kaye (with music by Saul Chaplin and lyrics by John Mercer). But it is most likely safe to say that no mathematical theorem has aroused as much interest among nonmathematicians as Gödel's incompleteness theorem, which appeared in 1931. The popular impact that this theorem has had in the last few decades can be seen on the Internet, where there are thousands of discussion groups dedicated to every topic under the sun. In any such group, it seems, somebody will sooner or later invoke Gödel's incompleteness theorem. One finds such invocations not only in discussion groups dedicated to logic, mathematics, computing, or philosophy, where one might expect them, but also in groups devoted to politics, religion, atheism, poetry, evolution, hip-hop, dating, and what have you. Interest in the incompleteness theorem is not confined to the Internet. In printed books and articles, we find the incompleteness theorem invoked or discussed not only by philosophers, mathematicians, and logicians, but by theologians, physicists, literary critics, photographers, architects, and others, and it has also inspired poetry and music.

Many references to the incompleteness theorem outside the field of formal logic are rather obviously nonsensical and appear to be based on gross misunderstandings or some process of free association. (For example, "Gödel's incompleteness theorem shows that it is not possible to prove that an objective reality exists," or "By Gödel's incompleteness theorem, all information is innately incomplete and self-referential," or again "By equating existence and consciousness, we can apply Gödel's incompleteness theorem to evolution.") Thus, Alan Sokal and Jean Bricmont, in their commentary on postmodernism [Sokal and Bricmont 98], remark that "Gödel's theorem is an inexhaustible source of intellectual abuses" and give examples from the writings of Regis Debray, Michel Serres, and others. But among the nonmathematical arguments, ideas, and reflections inspired by Gödel's theorem there are also many that by no means represent postmodernist excesses, but rather occur naturally to many people with very different backgrounds when they think about the theorem. Examples of such reflections are "there are truths that logic and mathematics are powerless to prove," "nothing can be known for sure," and "the human mind can do things that computers can not."

The aim of the present addition to the literature on Gödel's theorem is to set out the content, scope, and limits of the incompleteness theorem in such a way as to allow a reader with no knowledge of formal logic to form a sober and soundly based opinion of these various arguments and reflections invoking the theorem. To this end, a number of such commonly occurring arguments and reflections will be presented, in an attempt to counteract common misconceptions and clarify the philosophical issues. The formulations of these arguments and reflections, when not attributed to any specific author, are adaptations of statements found on the Internet, representative of many such reflections.

Gödel presented and proved his incompleteness theorem in an Austrian scientific journal in 1931. The title of his paper (written in German) was, translated, "On formally undecidable propositions of Principia Mathematica and related systems I." (A part II was planned, but never written.) *Principia Mathematica* (henceforth PM) was a monumental work in three volumes by Bertrand Russell and Alfred North Whitehead, published 1910–1913, putting forward a logical foundation for mathematics in the form of a (far from transparent) system of axioms and rules of reasoning within which all of the mathematics known at the time could be formulated and proved. Gödel proved two theorems in his paper, known as the first incompleteness theorem and the second incompleteness theorem. (The designation

"Gödel's incompleteness theorem" is used to refer to the conjunction of these two theorems, or to either separately.) The first incompleteness theorem established that on the assumption that the system of PM satisfies a property that Gödel named ω-consistency ("omega consistency"), it is *incomplete*, meaning that there is a statement in the language of the system that can be neither proved nor disproved in the system. Such a statement is said to be *undecidable* in the system. The second incompleteness theorem showed that if the system is *consistent*—meaning that there is no statement in the language of the system that can be *both* proved and disproved in the system—the consistency of the system cannot be established within the system.

The property of ω-consistency, which Gödel assumed in his proof, is a stronger property than consistency and has a technical flavor, unlike the more readily understandable notion of consistency. However, the American logician J. Barkley Rosser showed in 1936 that Gödel's theorem could be strengthened so that only the assumption of plain consistency was needed to conclude that the system is incomplete.

In the technical details, Gödel did not in fact carry through his argument for the system of PM, but rather for a system that he called P, related to that of PM. Nevertheless, it was clear that his result also applied to PM, and indeed to a wide range of axiomatic systems for mathematics, or parts of mathematics. Today the incompleteness theorem is often formulated as a theorem about any formal system within which a certain amount of elementary arithmetic can be expressed and some basic rules of arithmetic can be proved. The theorem states that any such system, if consistent, is incomplete, and the consistency of the system cannot be proved within the system itself.

The kind of reasoning put forward in Gödel's paper was at the time unfamiliar to logicians and mathematicians, and even some accomplished mathematicians (for example, the founder of axiomatic set theory, Ernst Zermelo) had difficulty grasping the proof. Today, as in the case of other intellectual advances, both the subject and our understanding of it have developed to the point where the proof is not at all considered difficult. The methods used have become commonplace, and proofs have become streamlined and generalized. Still, of course, grasping the details requires some familiarity with the methods and concepts of formal logic. In this book, no knowledge of logic or mathematics (beyond an acquaintance with basic school mathematics) will be assumed, on the basis of the view that a sound informal understanding of the theorem is attainable without a study of formal logic, and similarly for an understanding of the various applications and misapplications of the theorem.

Of course, opinions about what constitutes a sound informal understanding of the incompleteness theorem will vary, as illustrated by statements such as the following:

> Gödel's incompleteness theorem can be intuitively understood without a mathematical approach and proof: the incompleteness concept appears in clearly recognizable form in Zen Buddhism.

The incompleteness theorem is a theorem about the consistency and completeness of formal systems. "Consistent," "inconsistent," "complete," "incomplete," and "system" are words used not only as technical terms in logic, but in many different ways in ordinary language, so it is not surprising that Gödel's theorem has been associated with various ideas relating to incompleteness, systems, and consistency in some informal sense. As will be commented on at some length in Chapter 4, such associations usually have little or nothing to do with the content of the incompleteness theorem, and the kind of intuitive understanding of the theorem that one might derive from a study of Zen Buddhism is not at all what this book is about.

1.2 Gödel's Life and Work

Gödel has been variously described as German, Austrian, Czech, and American. All of these descriptions are correct in their own way. Gödel's language was German, as was his cultural background, but he was born into a well-to-do family in 1906 in the Moravian city of Brünn (Czech name Brno) in Central Europe, then a center for the textile industry. At the time of Gödel's birth, Moravia was part of the Austro-Hungarian empire. After World War I the Austro-Hungarian empire was dismantled, and Gödel grew up as a member of the sizeable German-speaking population of Brno and a citizen of the newly created state of Czechoslovakia. In 1929, when he was working on his doctoral dissertation at the University of Vienna, he became an Austrian citizen. After Austria was annexed by Nazi Germany in 1938, he was also, at least in the eyes of the authorities of that country, a German national. Although not Jewish and apparently completely unconcerned with politics, Gödel, as an academic and intellectual moving in Jewish and liberal circles, was viewed with some suspicion. He encountered difficulties in seeking an appointment as a lecturer and was once attacked in the street by a gang of Nazi youths (who were chased off by his wife). Such things, and the risk of conscription into the German army, prompted Gödel to seek to emigrate to the United States, which he managed to do after various complications in 1940, when he joined the Institute for Advanced Study (IAS) in Princeton. In 1948 he became an American citizen.

Although Gödel is famous mainly for his incompleteness theorem, he proved several other fundamental results in logic, chiefly while in Vienna between 1929 and 1940, during which time he also made visits to the IAS and to the University of Notre Dame in the United States. He proved, in his doctoral dissertation, the completeness theorem for first-order logic, which will be explained in Chapter 7, and went on to prove the first, and a little later the second, incompleteness theorem. He proved some highly significant results in set theory which will not be considered in this book, except for some incidental remarks. These were all seminal works which led to a number of developments in mathematical logic and the foundations of mathematics. He also proved other significant results during this very productive period. In the 1940s, as a member of the IAS where he remained until his retirement in 1976, he developed what is known as Gödel's Dialectica interpretation (after the journal in which this contribution first appeared, as late as 1958), having to do with what is called constructive mathematics. He also did original work in Einstein's general theory of relativity, demonstrating the existence of a solution of the equations of the theory that describes a universe in which it is theoretically possible to travel into one's past. Otherwise, his energies were chiefly directed toward philosophy. After 1940 he published little, but in the later years of his life he received numerous academic honors. He died in 1978.

Readers interested in further details about Gödel's life, such as his recurring health problems, mental and physical, his marriage, his life as an academic in Vienna, and his friendship with Einstein, should consult [Dawson 97] and the other sources listed in the references.

Large claims are often made about the impact and importance of the incompleteness theorem: for example, that it "brought a revolution to mathematical thought," that it "turned not only mathematics, but also the whole world of science on its head," and, in a crescendo of woolly enthusiasm, that Gödel's work "has revolutionized not only mathematics, but philosophy, linguistics, computer science, and even cosmology." Such claims are wildly exaggerated. Even in mathematics, where one might expect the incompleteness theorem to have had its greatest impact, it brought about no revolution whatsoever. The theorem is used all the time in mathematical logic, a comparatively small subfield of mathematics, but it plays no role in the work of mathematicians in general. To be sure, mathematicians are generally aware of the phenomenon of incompleteness and of the possibility of a particular problem being unsolvable within the standard axiomatic framework of mathematics, but a special case needs to be made in each instance where there is reason to believe that incompleteness should be a matter of mathematical concern. Gödel's "rotating universes," his new solutions of the equations of general relativity, have had no great impact on cosmology, and the subject of computer science could hardly have been revolutionized by Gödel's theorem since it didn't exist at the time the theorem was proved. The theoretical basis of computer science is associated rather with the work of the British mathematician and logician Alan Turing, who introduced in 1936 an idealized theoretical model of a digital computer and used it to prove the "unsolvability of the halting problem." This result is closely related to the first incompleteness theorem, and the basic connections between the two will be set out in later chapters.

Like the special theory of relativity a quarter of a century earlier, the theorems proved by Gödel in 1929 and 1930—the completeness theorem for first-order logic and the incompleteness theorem—were in the air at the time. Gödel himself felt that it would have been only a matter of months before somebody else had stumbled on the theorems ([Kreisel 80]). In the case of the completeness theorem, Gödel believed (rightly or wrongly) that only philosophical prejudice against nonfinitary reasoning had prevented the Norwegian logician Thoralf Skolem from arriving at the theorem. In the case of the first incompleteness theorem, priority was in fact claimed by the German mathematician Paul Finsler, but although his outlined argument can be made precise and correct using Gödel's work, it did not in his presentation amount to a proof of anything. The Polish-American logician Emil Post, who did pioneering work in the theory of computability, came much closer to Gödel's insight, but without producing any conclusive formal result. In particular, the mathematical precision and thoroughness of Gödel's proof of the first incompleteness theorem was probably necessary for the second incompleteness theorem to emerge as a corollary.

Apart from its manifold applications in logic, the incompleteness theorem does raise a number of philosophical questions concerning the nature of logic and mathematics. These questions, and the implications of the incompleteness theorem for our thinking about mathematics, are quite interesting and significant enough without any exaggerated claims for the revolutionary impact of the theorem. In particular, there are two philosophically highly significant aspects of the incompleteness theorem that will be touched on in this book. First, it shows that even very abstract mathematical principles, asserting the existence of various infinite sets, have formal consequences in elementary number theory that cannot be proved by elementary means. Secondly, when applied to formal systems whose axioms we recognize as mathematically valid, the incompleteness theorem shows that we cannot formally specify the sum total of our mathematical knowledge.

1.3 The Rest of the Book

The exposition adheres to the traditional plan of presenting general explanations before more specific discussions, and introducing and explaining concepts before they are used, but the reader is encouraged to dip into the book at any point of interest and to read other parts of the book or return to parts already visited when this appears profitable as a result of further reflection. The index will show where unfamiliar terms are first introduced and explained.

The incompleteness theorem is a mathematical theorem about axiomatizations of (parts of) mathematics, and the overview of the theorem in Chapter 2 accordingly begins with the subject of arithmetic. A reader who is chiefly interested in the supposed applications of the incompleteness theorem outside mathematics and the philosophy of mathematics may prefer to turn directly to Chapter 4 and later chapters, while a reader with an interest in mathematics will find in Chapters 2 and 3 an introduction to the mathematics of incompleteness, along with a discussion of some basic philosophical issues. Invocations of the incompleteness theorem in theology and in the philosophy of mind ("Lucas-Penrose arguments") are covered in Chapters 4, 5, and 6, and a discussion of the philosophical claims of Gregory Chaitin is found in Chapter 8. An Appendix has been added for the benefit of readers who are interested in a presentation of some of the formal mathematical aspects of incompleteness.

In order to make the book a bit more browsable, there is a certain amount of repetition of material from Chapter 2 in later parts of the book.

2

The Incompleteness Theorem An Overview

2.1 Arithmetic

The language of mathematics is full of terms and symbols that mean nothing to nonmathematicians, and fairly often indeed mean very little to anyone who is not an expert in a particular field of mathematics. But the part of mathematical language known in logic as the *language of elementary arithmetic* can be understood on the basis of ordinary school mathematics. It deals with the natural numbers (nonnegative integers) 0, 1, 2,... and the familiar operations of addition and multiplication, and it allows us to formulate some of the most striking results in mathematics, and some of the most difficult problems.

A *prime*, or prime number, is a natural number greater than 1 that is evenly divisible only by 1 and itself. Thus, the first few primes are 2 (the only even prime), 3, 5, 7, 11, 13, 17,.... One of the first substantial results of pure mathematics in the Western world was the discovery that the primes are inexhaustible, or infinite in number. In other words, for any given prime, there is a larger one. This was proved by a simple and ingenious argument in Euclid's *Elements* (ca. 300 B.C.). Some 2,000 years later, it was observed that if we look at the even numbers 0, 2, 4, 6, 8,..., it seems that beginning with 4 they can all be written as the sum of two primes: $4 = 2+2$, $6 = 3+3$, $8 = 5+3$, $10 = 5+5$, $12 = 7+5$,.... In this case, however, no proof suggested itself, and the statement in elementary arithmetic known as

Goldbach's conjecture, "Every even number greater than 2 is the sum of two primes," has not yet been either proved or disproved. Another conjecture about primes that has not yet been settled is the *twin prime conjecture*, according to which there are infinitely many primes p such that $p + 2$ is also a prime.

Let us take a closer look at a particular class of arithmetical problems. These problems are most conveniently described in terms of the *integers*, which besides the natural numbers also encompass the negative numbers $-1, -2, -3, \ldots$. A *Diophantine equation* (named after the third century Greek mathematician Diophantus) is an equation of the form $D(x_1, \ldots, x_n) = 0$, where $D(x_1, \ldots, x_n)$ is a polynomial with integer coefficients. What this means is that $D(x_1, \ldots, x_n)$ is an expression built up from the unknowns $x_1, \ldots, x_n$ using integers, multiplication, addition, and subtraction. A *solution* of the equation is an assignment of integer values to $x_1, \ldots, x_n$ such that the expression has the value 0. Some examples will make this clearer. The following (where we write x^2 for $x \times x$, y^4 for $y \times y \times y \times y$, $5y$ for $5 \times y$, and so on) are Diophantine equations:

$$x + 8 = 5y$$

$$x^2 = 2y^2$$

$$x^2 + y^2 = z^2$$

$$x^4 + y^4 = z^4$$

$$y^2 = 2x^4 - 1$$

$$x^{18} - y^{13} = 1.$$

The right-hand side of these equations is not 0, but we can easily rewrite the equations so as to get an equation in the form $D(x_1, \ldots, x_n) = 0$. The first equation becomes $x + 8 - 5y = 0$, the second equation $x^2 - 2y^2 = 0$, and so on.

The study of Diophantine equations—finding and describing their solutions, or determining that they have no solutions—has been a specialized field of mathematics for centuries. Diophantine problems range from the very simple to the apparently hopelessly difficult, and mathematicians have displayed extraordinary ingenuity in studying various classes of Diophantine equations.

The first of the equations presented as examples clearly has infinitely many solutions: for any integer n we get a solution with $y = n$ by setting $x = 5n - 8$. The second equation has no other solution than $x = 0$, $y = 0$. This was also proved in the *Elements*, where it is shown by an argument traditionally attributed to the school of Pythagoras that 2 has no rational square root. By the theorem of Pythagoras about the square of the hypotenuse, this is the same as saying that the diagonal of a square is incommensurable with the side of the square. (In other words, there is no unit of length such that the side is measured by n units and the diagonal by m units for some natural numbers n and m.) The third equation has infinitely many solutions, which are called *Pythagorean triplets* (because of the connection with the theorem of Pythagoras). The particular solution $x = 3$, $y = 4$, $z = 5$ was used in antiquity as a means of obtaining right angles in practical surveying. The fourth equation has no solution in nonzero integers, although proving this is far from trivial. Fermat's Last Theorem (really Fermat's claim or conjecture), which was finally proved in 1994 through the combined efforts in advanced mathematics of several mathematicians, states that no equation of the form $x^n + y^n = z^n$ with n greater than 2 has any solution in positive integers. The fifth equation has only two positive solutions: $x = 1$, $y = 1$, and $x = 13$, $y = 239$. This was proved (using some very complicated mathematical reasoning) by the Norwegian mathematician Wilhelm Ljunggren in 1942. The last equation has no solution where x and y are both nonzero. This is a consequence of *Catalan's conjecture* (1844), which states that 8, 9 is the only pair of two consecutive natural numbers where each is a perfect power, that is, equal to n^k for some n and k greater than 1. Catalan's conjecture was finally proved in 2002 by Preda Mihailescu.

Problems in arithmetic do not always involve primes or Diophantine equations. The *Collatz conjecture* (also known by other names, for example the $3n+1$ conjecture and Ulam's problem) states that if we start with any positive natural number n and compute $n/2$ if n is even, or $3n + 1$ if n is odd, and continue applying the same rule to the new number, we will eventually reach 1. For example, beginning with 7, we get

$$7, 22, 11, 34, 17, 52, 26, 13, 40, 20, 10, 5, 16, 8, 4, 2, 1.$$

No proof of the Collatz conjecture has been found in spite of intense efforts, and the problem is generally considered extremely difficult.

A Logical Distinction

The subject of arithmetic plays a large role in a discussion of the incompleteness theorem, but fortunately we don't need to tackle any difficult arithmetical problems. Instead, we will be concerned with some relatively simple *logical* aspects of such problems.

Goldbach's conjecture states that every even number greater than 2 is the sum of two primes. In an equivalent formulation, Goldbach's conjecture states that every natural number has the property of being smaller than 3 or odd or the sum of two primes. The point of this reformulation is that there is an *algorithm* for deciding whether or not a given number has this property. An algorithm is a purely mechanical, computational procedure, one that when applied to a given number or finite sequence of numbers always terminates, yielding some information about the numbers. For example, in school we learn algorithms that given two numbers n and k yield as output their sum $n + k$ and their product $n \times k$. We also have algorithms for comparing two given numbers, and for dividing a natural number by a nonzero natural number, yielding a quotient and remainder. Combining these algorithms, we get an algorithm for deciding whether a given number has the property of being smaller than 3 or odd or the sum of two primes. For if n is the sum of two primes, these must both be smaller than n, so to check whether an even number n greater than 2 is the sum of two primes, we go through the numbers smaller than n, looking for two primes that add up to n. Whether a number n is a prime is again something that can be decided by a computation—just divide n by the numbers 2, $3,\ldots,n-1$, and check whether the remainder is 0 in any division.

A property of numbers that can be checked by applying an algorithm is called a *computable* property. (This notion will be explored further in Chapter 3.) As we have seen, Goldbach's conjecture can be formulated as a statement of the form "Every natural number has the property P," where P is a computable property. This is a logically very significant feature of Goldbach's conjecture, and in the following, any statement of this form will be called a *Goldbach-like* statement. (In logic, these are known by the more imposing designation "Π-0-1-statements.") Actually, this description glosses over an important point: the property P must not only be computable, but must also have a sufficiently simple form so that an algorithm for checking whether a number has the property P can be "read off" from the definition of P. It will be clear in the case of all examples of Goldbach-like statements considered in the book that this condition is satisfied. (A formal definition of the Goldbach-like statements in the language of arithmetic is given in the Appendix.)

We extend the class of Goldbach-like statements to statements of the form "Every finite sequence $k_1, \ldots, k_n$ of natural numbers has the property P," where P again is a computable property, so that there is an algorithm which given any sequence of numbers $k_1, \ldots, k_n$ decides whether or not it has property P.

If P is a computable property, then so is the property not-P. So every statement of the form "There is no natural number k with property P," where P is a computable property, can be equivalently formulated as a Goldbach-like statement: "Every natural number has the property not-P."

A *counterexample* to the statement "Every natural number has property P" is a natural number which does *not* have property P. A consequence of a statement being Goldbach-like is that if it is false, it can be *disproved* very simply. To disprove it, we need only carry out a computation showing that some number n is in fact a counterexample and conclude that the statement is false. Of course, if the shortest such computation is extremely lengthy, "can be disproved" here only means "can in principle be disproved." In other words, we need to disregard all limitations of time, space, and energy. But at any rate, we can observe that if the statement is false, it is a *logical consequence* of the basic rules of arithmetic that it is false, in the sense that a lengthy computation using those rules, if possible to carry out, would show it to be false. In other words, there exists, in the mathematical sense, a formal proof that the statement is false, one that uses only the basic rules of arithmetic. Here we are regarding computations as a special kind of proof, in the sense of a mathematical argument showing some statement to be true. In logic, we study *formal systems*, axiomatic theories in which mathematical statements can be proved or disproved, and the incompleteness theorem is a theorem about those formal systems within which arithmetical computations can be carried out. More will be said about formal systems, beginning in Section 2.2.

It was noted above that a statement of the form "Every natural number has property P," where P is a computable property, can always in principle be disproved if it is false by exhibiting a counterexample and carrying out a computation. A further significant observation is that a counterexample, if it exists, can always be *found* by just systematically checking 0, 1, 2, 3,... until we get to the smallest number that does not have the property P. Thus, there is also a systematic procedure that will (in principle) eventually *find* a disproof of a Goldbach-like statement if it is false. This is a special case of the general fact that for any formal system S, if a statement A is provable in S, a systematic search will eventually find such a proof of A in S. If A is not provable in S, a systematic search will in general just go on forever without yielding any result.

So for any formal system S that incorporates a bit of arithmetic—the basic rules needed to carry out computations—a Goldbach-like statement is disprovable in S if false. On the other hand, we cannot make any similar observations about how a Goldbach-like statement can be *proved* if it is true. For every n, a computation can indeed verify that every number $0, 1, \ldots, n$ has the property P, but this is not a verification that *every* number has property P, no matter how large n is chosen. If a Goldbach-like statement is true, it may well be that it can be proved to be true, but we cannot say at the outset what mathematical methods such a proof might require.

Every statement of the form "The Diophantine equation $D(x_1, \ldots, x_n) = 0$ has no solution in nonnegative integers" is a Goldbach-like statement. Here the relevant property of a sequence of numbers $k_1, \ldots, k_n$ is that of not being a solution of the equation $D(x_1, \ldots, x_n) = 0$, and checking this property only involves carrying out a series of multiplications, additions, and subtractions, to see whether the result is 0.

In contrast, the twin prime conjecture is *not* a Goldbach-like statement. It can be expressed as "Every natural number has the property P," where a natural number n has the property P if there is a prime p larger than n such that $p+2$ is also a prime. But in this case we cannot read off from the definition of the property any algorithm for checking whether a number has this property or not. The procedure of systematically looking for a pair of primes p and $p + 2$ greater than n will never terminate if there is no such prime, and so it can never give the answer that n does not have the property. Of course, if the twin prime conjecture is true, the procedure is in fact an algorithm, one that will always show n to have the property, but as long as we don't know whether the conjecture is true or not, we don't know whether the procedure is an algorithm. Similar remarks apply to the Collatz conjecture. It states that a particular sequence of computations will always (that is, for every starting number) lead to 1, but there is no obvious algorithm for deciding whether a particular starting number does lead to 1. In the case of these two conjectures, we therefore have no logical grounds for claiming that they must be disprovable if false.

The property of an arithmetical statement of being Goldbach-like will play a role at several points in the discussion of incompleteness.

2.2 The First Incompleteness Theorem

As noted in Section 2.1, although the problems of elementary arithmetic are often easily *stated*, there is no limit on the difficulty or complexity of the mathematics needed to *solve* those problems. Thus, the 129-page proof by Andrew Wiles of Fermat's theorem was in fact a proof of what is known as the Taniyama-Shimura conjecture for elliptic curves in the semi-stable case, which had earlier been shown by K. A. Ribet to imply Fermat's theorem. We don't need to know what any of this means in order to appreciate the basic point that Fermat's theorem, although easily stated in elementary mathematical language, was proved using a detour into very complicated and advanced mathematics.

Since Fermat claimed to have a proof of his conjecture, one which unfortunately didn't fit into the margin of the work by Diophantus that he was reading, many have sought (and sometimes mistakenly believed that they have found) an *elementary* proof of the theorem, in the sense of a proof that does not make use of any mathematics unknown to Fermat and his contemporaries. However, there is little doubt that Fermat was himself mistaken in thinking that he had a proof of the theorem. It is an open question to what extent the existing proof of Fermat's last theorem can be simplified, and it remains to be seen if any essentially simpler proof can be extracted from that proof. (There is some reason to believe, on general grounds, that an elementary proof of the theorem exists in a purely theoretical sense, a proof that would not only fail to fit into the margin of Fermat's book, but would if printed out require thousands of pages.)

In view of the 300 years of mathematical development needed for mathematicians to finally solve the problem posed by Fermat's claim, and the fact that Goldbach's and other arithmetical conjectures have not yet been settled, one may wonder whether there is any guarantee that all arithmetical problems posed by mathematicians *can* be solved, given sufficient time and effort. A conviction that this is the case was expressed by the German mathematician David Hilbert in a famous address at the international congress of mathematicians in Paris in the year 1900, where he presented 23 major problems facing mathematicians in the new century:

> Take any definite unsolved problem, such as the question as to the irrationality of the Euler-Mascheroni constant C, or the existence of an infinite number of prime numbers of the form $2^n + 1$. However unapproachable these problems may seem to

> us, and however helpless we stand before them, we have, nevertheless, the firm conviction that their solution must follow by a finite number of purely logical processes. ... This conviction of the solvability of every mathematical problem is a powerful incentive to the worker. We hear within us the perpetual call: There is the problem. Seek its solution. You can find it by pure reason, for in mathematics there is no *ignorabimus.*

This is one expression of Hilbert's optimistic "non ignorabimus." Hilbert was alluding to an old saying, "ignoramus et ignorabimus" (we do not know and we shall never know), which the physiologist Emil du Bois-Reymond had affirmed in 1872, speaking of our knowledge of human consciousness and the physical world.

Gödel's first incompleteness theorem by no means *refutes* this optimistic view of Hilbert's. What it does is establish that Hilbert's optimism cannot be justified by exhibiting any single formal system within which all mathematical problems are solvable, even if we restrict ourselves to arithmetical problems:

First incompleteness theorem (Gödel-Rosser). *Any consistent formal system S within which a certain amount of elementary arithmetic can be carried out is incomplete with regard to statements of elementary arithmetic: there are such statements which can neither be proved, nor disproved in S.*

To understand what this means, we need to start by considering the notion of a formal system. This will be followed by some comments on the consistency requirement and then on the condition of encompassing "a certain amount of arithmetic."

Formal Systems

A formal system is a system of *axioms* (expressed in some formally defined language) and *rules of reasoning* (also called *inference rules*), used to derive the *theorems* of the system. A theorem is any statement in the language of the system obtainable by a series of applications of the rules of reasoning, starting from the axioms. A *proof* in the system is a finite sequence of such applications, leading to a theorem as its conclusion.

The idea of an axiomatic system of this kind is an old one in mathematics and has in the past been familiar to many through a study of Euclidean geometry. Euclid, in the *Elements*, introduces a number of *definitions*, such as Definition 1, "A point is that which has no part." He further introduces *postulates*, which are basic principles of geometry (the most famous of these being the parallel postulate), and *common notions*, which have the character of general rules of reasoning. An example of a common notion is "Things which equal the same thing also equal one another." Using the definitions, postulates, and common notions, Euclid derives a large number of theorems (*propositions*).

Euclid's definitions, postulates, and common notions do not amount to a formal system. The language of the system is not formally specified, his proofs use geometrical assumptions not expressed in the postulates, and they use other logical principles than those expressed in the common notions. (Formal axiomatizations of geometry were given only in the twentieth century.) But Euclid's geometry was the basic model for the axiomatic method for millennia, and the idea of organizing knowledge through axioms, definitions, and rules of reasoning leading to theorems has been vastly influential in philosophy, science, and other fields of thought. However, it is only in mathematics that the axiomatic method has been strikingly successful in the organization and analysis of knowledge.

In modern logic, many specific formal axiomatic systems are studied, often of a kind called *first-order theories*. Here the word "theory" is logical jargon and does not have any of the connotations often associated with the word in everyday or scientific contexts—a theory in the sense used in logic is just an axiomatic formal system. In this book, "theory" will often be used as a synonym of "formal system." "First-order" refers to a particular collection of rules of reasoning used in proving theorems, namely the rules for which Gödel proved his (confusingly named) *completeness theorem*, which will be commented on in Section 2.3.

Two first-order theories prominent in logic to which the incompleteness theorem applies are Peano Arithmetic, or PA, which is a formal theory of elementary arithmetic, and Zermelo-Fraenkel set theory with the axiom of choice, ZFC. (The designation ZF is used for the same theory without the axiom of choice.) In this book, we will not study these theories (although the axioms of PA are described in Section 7.2), but some general observations about them will be relevant to the argument of the book. At this point we need only observe that PA is a formal system within which all of the arithmetical reasoning that is usually described as *elementary* can be carried out, while ZFC is an extremely powerful system that suffices for formally proving most of the theorems of present day mathematics.

Negation

The incompleteness theorem applies to many other kinds of formal systems than first-order theories, but we will assume that the language of a formal system at least includes a *negation operator*, so that every sentence A in the language has a negation not-A. (A sentence is here a full sentence, expressing a statement which it makes sense to speak of as proved or disproved.) This allows us to define what it means for S to be *consistent* (there is no A such that both A and not-A are theorems) and for a sentence A to be *undecidable* in S (neither A nor not-A is a theorem of S). A system is *complete* (sometimes called "negation complete") if no sentence in the language of S is undecidable in S, and otherwise *incomplete.*

We need to introduce a bit of notation: if A is a sentence in the language of S, we denote by $S+A$ the formal system obtained by adding A as a new axiom to the axioms of S (but not changing the rules of inference). If A is provable in S, $S+A$ has the same theorems as S (with A unnecessarily taken as an axiom), while if A is not provable in S, $S+A$ is a stronger theory, which in particular proves A.

Here, we encounter a point of usage where logical terminology is somewhat at odds with ordinary usage. It may seem odd to say that A, which is an axiom of $S+A$, is *provable* in $S+A$. In ordinary informal contexts, it is often taken for granted that axioms are not provable, but are basic assumptions that cannot be proved. In logic, however, "provable" is relative to some theory, and every axiom of a theory S is also provable in S. The proof of an axiom A is a trivial one-liner, in which it is pointed out that A is an axiom. In mathematics, we do not usually say that axioms are provable, but we do say that statements follow "immediately" or "trivially" from the axioms. For example, from the axiom "for every n, $n+0=n$" it follows immediately that $0+0=0$. In logic we make no such fine distinction between axioms and immediate consequences of axioms—they are all provable.

We can now state a couple of basic connections between negation, provability, consistency, and undecidability. The first connection consists in the fact that an inconsistent theory has no undecidable statements. This is because in an inconsistent theory, *every* statement in the language of the theory is provable, by a rule of reasoning known in logic by the traditional name of *ex falso quodlibet*, "anything follows from a falsehood." (It should really be "anything follows from a contradiction.")

The second connection is expressed in the following observations:

> A is provable in S if and only if $S +$ not-A is inconsistent, and A is undecidable in S if and only if both $S + A$ and $S +$ not-A are consistent.

These observations will be used frequently in the following. They depend on the rule that not-not-A is logically equivalent to A, although the incompleteness theorem can also be adapted to systems in which this rule does not hold. To verify the observations, we use the fact that a proof of a statement B in the theory $T + A$ can also be regarded as an argument in T leading from the assumption A to the conclusion B, and thus as a proof of "if A then B" in T. Conversely, if "if A then B" is a theorem of T, B is a theorem in the theory obtained by adding A as an axiom. Specializing this to a logical contradiction B (a statement of the form "C and not-C") yields the stated observations.

Formal Systems and the Theory of Computability

The incompleteness theorem applies not only to systems like PA and ZF, which formulate part of our mathematical knowledge, but also to a wide class of formal systems, those in which a "certain amount of arithmetic" can be carried out, whether or not their axioms express any part of our mathematical knowledge and even if the axioms are false or not associated with any interpretation at all.

In everyday usage it would perhaps be odd to speak of false axioms, or axioms not associated with any interpretation. In logic, however, when we speak of formal systems in general, the word "axiom" must not be assumed to be reserved for statements that are in some sense basic and irreducible to simpler principles, or statements that we believe to be true or that are in some other sense acceptable. Instead, the general concept of an axiom in logic is strictly relative to a formal system, and any sentence A in a formal language can be chosen as an axiom in a formal system.

This very general notion of a formal system is closely associated with the theory of computable properties. The reason for this is that when we try to characterize the general concept of a system of axioms and rules of reasoning, we need to impose some condition to the effect that recognizing an axiom or applying a rule must be a mechanical matter—we shouldn't have to solve further mathematical problems in order to be able to decide whether an application of a rule is correct or not, or whether a particular statement is in fact an axiom. Otherwise, we would need to introduce a second system of axioms and rules of reasoning for proving that something is in fact a proof in the first system. So it is required of a formal system that in order to verify that something is an axiom or an application of a rule of reasoning, we do not need to use any further mathematical reasoning, but need only apply mechanical checking of a kind that can be carried out by a computer.

In popular formulations of Gödel's theorem, a condition of this kind (as far as the axioms are concerned) is sometimes included in the form of a stipulation that the axioms of a formal system are *finite* in number. This implies that an axiom can ("in principle") be recognized as such by looking through a finite table. But this condition is not in fact satisfied by many of the formal systems studied in logic, such as PA and ZF. These systems have an infinite number of axioms, but it is still a mechanical matter to check whether or not a particular sentence is an axiom. For example, PA incorporates the rule of reasoning known as "proof by mathematical induction" by having an infinite number of axioms of the form "if 0 has property P, and $n+1$ has property P whenever n has property P, then every number has property P." We can recognize every instance of this principle by a simple inspection, even though there are infinitely many possible choices of P. Another example of a theory with infinitely many axioms, where the axioms are not just instances of a general schema as in PA, will be given in Section 5.4.

Thus, in a general characterization of formal systems, we need to make use of the general notion of a mechanically computable property, for which a theoretical foundation was given by the work of Turing and others in the 1930s. The notion of a computable property was introduced earlier in connection with Goldbach-like statements when applied to properties of numbers, but it applies equally to properties of sentences and finite sequences of sentences in a formal language. The concept of computability, and its connection with formal systems and with the incompleteness theorem, will be treated in a more systematic fashion in Chapter 3.

Consistency

The first incompleteness theorem, in the form given (which incorporates Rosser's strengthening of Gödel's result), only assumes that the system

S, apart from incorporating a certain amount of arithmetic, is *consistent.* This is a strong result because consistency is not a very strong condition to impose on a theory. Consistent theories of arithmetic, like consistent liars, can spin a partly false and completely misleading (mathematical) yarn.

Suppose we know that ZFC proves an arithmetical statement A. Can we conclude that the problem of the truth or falsity of A is thereby solved? If we have inspected the proof, we will regard the problem as solved if we find this particular proof convincing, even if we have doubts about the axioms of ZFC in general. If all we know is that a proof in ZFC *exists*, we will accept the problem as solved if we have confidence in the theory as a whole. But mere belief in the consistency of ZFC is in general insufficient to justify accepting A as true on the basis of the knowledge that it is provable in ZFC.

There is a class of statements that are guaranteed to be true if provable in a consistent system S incorporating some basic arithmetic. Suppose A is a *Goldbach-like* statement. We can then observe that if A is provable in such a system S, it is in fact true. For if A is false, it is provable in S that A is false, since this can be shown by a computation applied to a counterexample, and so if S is consistent, it cannot also be provable in S that A is true. Thus, for example, it is sufficient to know that Fermat's theorem is provable in ZFC and that ZFC is consistent to conclude that the theorem is true. But in the case of a statement that is *not* Goldbach-like, for example the twin prime conjecture, we cannot in general conclude anything about the truth or falsity of the conjecture if all we know is that it is provable, or disprovable, in some consistent theory incorporating basic arithmetic.

The incompleteness theorem gives us concrete examples of consistent theories that prove false theorems. This is most easily illustrated using the second incompleteness theorem. Given that ZFC is consistent, ZFC + "ZFC is inconsistent" is also consistent, since the consistency of ZFC is not provable in ZFC itself, but this theory disproves the true Goldbach-like statement "ZFC is consistent." (That this statement is in fact Goldbach-like will be seen in Section 2.6.)

In logic, one speaks of *soundness properties* of theories. When we are talking about arithmetic, consistency is a minimal soundness property, while the strongest soundness property of a theory is that every arithmetical theorem of the theory is in fact true. Among the intermediate soundness properties, a particularly prominent one is that of not *disproving* any *true* Goldbach-like statement. In the logical literature, this property is sometimes called "1-consistency," sometimes "Σ-soundness" (Sigma soundness), and in this book the latter term will sometimes be used. (The reason for this terminology will emerge in the Appendix.) The property of ω-consistency, which Gödel used in the formulation of his theorem, implies Σ-soundness, but is a stronger property.

Note that since a false Goldbach-like statement is disprovable in any theory encompassing basic arithmetic, if a Goldbach-like statement is *undecidable* in such a theory (which must be consistent, since it has undecidable statements) it is also true. Thus, a *proof* of the undecidability of Goldbach's conjecture in PA would at the same time be a proof of the conjecture. There is nothing impossible about such a situation, since the proof of undecidability could be carried out in a stronger theory such as ZFC. But it is also conceivable that we could prove a hypothetical statement such as "If Goldbach's conjecture is true, then it is unprovable in ZFC" without ever deciding Goldbach's conjecture itself. There are Goldbach-like statements A for which we can indeed prove such hypothetical statements without being able to prove or disprove A. This also follows from the second incompleteness theorem: even if we have no idea whether or not S is consistent, we can prove the hypothetical statement "if S is consistent, the consistency of S is unprovable in S."

The "Certain Amount of Arithmetic"

The first incompleteness theorem applies to formal systems within which a certain amount of elementary arithmetic can be carried out. We need to be a bit more explicit about what this means.

Any system whose language *includes* the language of elementary arithmetic, and whose theorems include some basic facts about the natural numbers, is certainly one that satisfies the condition. (A sufficient set of "basic facts" of arithmetic is specified in the Appendix.) But the incompleteness theorem also applies to systems that do not make any explicit statements about natural numbers, but instead refer to mathematical objects that can be used to *represent* the natural numbers. For example, strings of symbols or certain finite sets can be used in such a representation. This notion of one kind of mathematical objects representing another kind will be further commented on in connection with the second incompleteness theorem. Here we need only note that nothing essential will be lost if we think of the formal systems to which the first incompleteness theorem applies as those systems that have an *arithmetical component*, in which we can use the language of arithmetic and establish some basic facts about addition and multiplication of natural numbers. Depending on the system, there may be some translation involved in using the language of arithmetic within the system, but essentially we can think of the language of the system as including the language of elementary arithmetic.

The essential point of the requirement of encompassing a "certain amount of arithmetic" can be explained without formally specifying the requisite amount. In the discussion of Goldbach-like statements, it was claimed that if a property of natural numbers, such as being the sum of two primes, can be checked by a mechanical computation, then if a number n has that property, there is an elementary mathematical proof that n has the property. The "certain amount of arithmetic" that a formal system S needs to encompass for the proof of the first incompleteness theorem to apply to S is precisely the arithmetic needed to substantiate this claim. In other words, if the "certain amount of arithmetic" can be carried out within S, S can prove all arithmetical statements that can established by means of a more or less lengthy mechanical computation.

The incompleteness theorem is often misstated as applying to systems that are "sufficiently complex." This is incorrect because the condition of encompassing a certain amount of elementary arithmetic does not turn on complexity in either a formal or informal sense, but on what can be *expressed* and what can be *proved* in a system. There are very simple systems to which the incompleteness theorem applies, and very complex ones to which it does not apply. The relation between complexity and incompleteness will be considered further in Chapter 8.

The idea of complexity also appears in supposed applications of the incompleteness theorem outside mathematics. The following is a quotation from "Postmodernism and the future of traditional photography" by Richard Garrod:

> Early in the 20th century, the mathematician Gödel established that in a system of sufficient complexity (and that level is reached in the syntax of any toddler) a complete description of that system is not possible. The complexity of the simplest photograph—any photograph—is incalculable, and the creative possibilities of the simplest photograph are in fact infinite—and always will be.

These comments are typical of many references to Gödel's theorem that loosely associate it with incompleteness or complexity in some sense or other. It may well be that the complexity of the simplest photograph is incalculable, but Gödel did not in fact establish that a "complete description of a system" is impossible if the system is "sufficiently complex." In the quotation, a photograph is apparently regarded as a "system," and whatever is intended by this, it seems clear that the author is not actually suggesting that a photograph proves any arithmetical statements. Thus, this is one of many references to Gödel's theorem that do not make contact with the actual content of the theorem, but rather invoke it as an analogy or metaphor, or as a general source of inspiration.

2.3 Some Limitations of the First Incompleteness Theorem

"Unprovable Truths"

It is often said that Gödel demonstrated that there are truths that cannot be proved. This is incorrect, for there is nothing in the incompleteness theorem that tells us what might be meant by "cannot be proved" in an absolute sense. "Unprovable," in the context of the incompleteness theorem, means unprovable in some particular formal system. For any statement A unprovable in a particular formal system S, there are, trivially, other formal systems in which A is provable. In particular, A is provable in the theory $S + A$ in which it is taken as an axiom.

Of course the trivial observation that A is provable in $S + A$ is of no interest if we're wondering whether A can be proved in the sense of being shown to be true. The idea of Gödel's theorem showing the existence of "unprovable truths" derives whatever support it has from the fact that there are formal systems in which we have included axioms and rules of reasoning that are correct from a mathematical point of view, and furthermore suffice for the derivation of *all* of our ordinary arithmetical theorems. In particular, ZFC is a very strong theory within which all of the arithmetical theorems of present-day mathematics are thought to be provable. So, since ZFC is incomplete with respect to arithmetical statements (given that the theory is consistent), can we conclude that there are arithmetical truths that are unprovable in the sense that there is no way for us, as human mathematicians, to mathematically establish the truth of the statement?

This conclusion would perhaps follow if we didn't know of any way of extending the axioms of ZFC to a stronger set of axioms which can still, with the same justification as the axioms of ZFC, be held to express valid principles of mathematical reasoning. But in fact there are such ways of extending ZFC (see Sections 5.4 and 8.3).

There is therefore nothing in the incompleteness theorem to show that there are true arithmetical statements that are unprovable in any absolute sense. Still, if an arithmetical sentence is unprovable in ZFC, there are good grounds for thinking that it cannot be proved using today's "ordinary" mathematical methods and axioms, such as one finds in mathematical textbooks, and also that it cannot be proved in a way that a large majority of mathematicians would today regard as unproblematic and conclusive.

Complete Formal Systems

The incompleteness theorem does not imply that *every* consistent formal system is incomplete. On the contrary, there are many complete and consistent formal systems. A particularly interesting example from a mathematical point of view is the elementary theory of the real numbers. (Another example, a theory known as Presburger arithmetic, is given in Chapter 7.) The language of this theory, like the language of arithmetic, allows us to speak about addition and multiplication of numbers, but now with reference to the *real* numbers rather than the natural numbers. The real numbers encompass the integers, but also all rational numbers m/n where m and n are integers, along with irrational numbers like the square root of 2 and the number π.

An example of a statement in this language is the following, taken from the Canadian mathematical olympiad of 1984:

> For any seven different real numbers, there are among them two numbers x and y such that $x - y$ divided by $1 + xy$ is greater than 0 and smaller than the square root of three.

Another such statement is

> An equation $x^3+bx^2+cx+d = 0$ has two different real solutions if and only if $3c - b^2$ is smaller than 0 and $4c^3 - b^2c^2 - 18bcd - 4b^3d + 27d^2$ is smaller than or equal to 0.

The complete theory of the real numbers proves these and similar statements. As should be clear from these examples, the theory is far from trivial, and it has numerous applications in electrical engineering, computational geometry, optimization, and other fields.

Since the natural numbers form a subset of the real numbers, it may seem odd that the theory of the real numbers can be complete when the theory of the natural numbers is incomplete. The incompleteness of the theory of the natural numbers does not carry over to the theory of the real numbers because even though every natural number is also a real number, we cannot *define* the natural numbers as a subset of the real numbers using only the language of the theory of the real numbers, and therefore we cannot express arithmetical statements in the language of the theory. Thus, we cannot, for example, in the theory of the real numbers express the statement "There are natural numbers m, n, k greater than 0 such that $m^3 + n^3 = k^3$." We *can* express the statement "There are real numbers r, s, t greater than 0 such that $r^3 + s^3 = t^3$" and also easily prove this statement.

How would we ordinarily define the natural numbers as a subset of the real numbers? The real numbers 0 and 1 can be identified with the corresponding natural numbers, and using addition of real numbers we get the natural numbers as the subset of the real numbers containing 0, 1, $1+1$, $1+1+1$, and so on. However, this "and so on" cannot be expressed in the language of the theory of real numbers. Another way of formulating this definition of the natural numbers is to say that the natural numbers are the real numbers that belong to every set A of real numbers that contains 0 and is closed under the operation of adding 1; that is, for which $x+1$ is in A whenever x is in A. This definition uses a *second-order* language, in which one can refer to *sets* of real numbers. The language of the elementary theory of the real numbers, like the language of elementary arithmetic, only allows us to refer to numbers, not to sets of numbers.

The Completeness Theorem

Gödel's first major work in logic was his proof that first-order predicate logic is complete. The statement of this theorem carries the unfortunate suggestion that predicate logic is a complete formal system in the sense of the incompleteness theorem, and comments such as the following are often encountered:

> First-order predicate logic is still not powerful enough to succumb to Gödel's incompleteness theorems. In fact Gödel himself proved it consistent and complete.

> Gödel is also responsible for proving (1930) that first-order predicate logic *is* complete. The incompleteness proofs apply only to formal systems strong enough to represent the truths of arithmetic.

Such comments are based on a natural and widespread misunderstanding caused by the fact that "complete" is used in logic in two different senses. That predicate logic is complete does not mean that some formal system is complete in the sense of Gödel's incompleteness theorem, or in other words, negation complete. Completeness in the context of the completeness theorem has a different meaning—that predicate logic is complete means that the rules of reasoning used in predicate logic are sufficient to derive every logical consequence of a set of axioms in a first-order language. Thus, a less misleading description of Gödel's early work would be to say that he proved that many formal systems are negation incomplete and also proved the adequacy for logical deduction of the inference rules of first-order logic.

The completeness theorem, and its consequences for incompleteness in the sense of Gödel's incompleteness theorem, will be commented on further in Chapter 7.

Undecidable Nonarithmetical Statements

A formal system S that contains other kinds of statements than arithmetical statements can of course have many undecidable nonarithmetical statements. What the incompleteness theorem shows is that S will be incomplete in its *arithmetical* component. In other words, there is a statement of arithmetic that cannot be decided in S. Furthermore, by putting enough work into the proof of the incompleteness theorem we can explicitly construct, given a specification of S, a particular Goldbach-like arithmetical statement that is undecidable in S.

Thus, the incompleteness theorem pinpoints a specific incompleteness in any formal system that encompasses some basic arithmetic: it does not decide every arithmetical statement. Unfortunately for the applicability of the incompleteness theorem outside mathematics, this also means that we learn nothing from the incompleteness theorem about the completeness or incompleteness of formal systems with regard to nonarithmetical or nonmathematical statements.

A weaker formulation of the first incompleteness theorem than that given states only that for any formal system that encompasses a certain amount of arithmetic, there is a statement in the language of the system which is undecidable in the system. This weaker formulation in a way sounds more interesting than the stronger one, since it suggests that a theory of astrophysics or ghosts that includes a bit of arithmetic cannot tell the whole story about astrophysics or ghosts. But this suggestion is unjustified. A consistent formal system may, for all we learn from Gödel's incompleteness theorem, be complete as regards statements about astrophysics, about ghosts or angels, about the human soul, about the physical universe, or about the course of the past or the future. The incompleteness theorem only tells us that the system cannot be complete in its arithmetical part.

Interesting-sounding, supposed applications of the incompleteness theorem outside mathematics, therefore, often ignore the essential condition of encompassing some basic arithmetic, and formulate the theorem incorrectly as a theorem about formal systems in general. When they do take this condition into account, authors may resort to such formulations as "Our hypothesis is that the universe is at least as big as arithmetic, so that it is affected by incompleteness" or "The philosophy should incorporate arithmetic, or else it is already limited." The first of these makes no apparent sense, while the second amounts to the feeble criticism of a system of philosophy that it cannot be a complete guide to life or the universe since it fails to decide every arithmetical statement. Such invocations of the incompleteness theorem will be further commented on in Chapter 4.

2.4 The First Incompleteness Theorem and Mathematical Truth

Truth and Undecidability

The words "sentence" and "statement" are used interchangeably in this book, although "sentence" is sometimes used specifically to emphasize the purely *syntactic* character of some of the concepts introduced. A purely syntactic concept is one that makes no reference to the meaning or interpretation of a language, or the truth or falsity of statements in the language, but only refers to formal rules for constructing sentences or proofs. For example, that a system S is incomplete means that there is some sentence A in the language of S such that neither A nor not-A is provable in S. This notion of incompleteness does not presuppose any notion of *truth* for the sentences in the language of S, but only refers to sequences of symbols formed according to certain rules (the sentences of the system) and sequences of sentences connected by applications of certain formal rules (the proofs in the system). Incompleteness is thus a purely syntactic concept.

The incompleteness theorem is often popularly formulated as saying that for systems S to which the theorem applies, there is some *true* statement in the language of S that is undecidable in S. If we do interpret the sentences in the language of S as expressing true or false statements, this is indeed a consequence of the incompleteness theorem, for if A is undecidable in S, then so is not-A, and one of A and not-A is true. But the incompleteness theorem applies even if we regard the sentences of S as mere sequences of symbols, not as expressing statements for which it makes sense to ask if they are true or not. For example, there are those who for philosophical reasons would hold that sentences in the language of ZFC do not in general express any true or false statements, but this does not prevent them from recognizing that the incompleteness theorem applies to ZFC.

This said, it also needs to be emphasized that since the condition for the incompleteness theorem to apply to a formal system is that it encompasses "a certain amount of arithmetic," we can always, when applying the incompleteness theorem to a system S, specify at least a *subset* of the sentences of S that we can interpret as expressing statements of arithmetic, and therefore as being true or false. (Although very different interpretations of these sentences may well be intended by the proposers of the system, as will be explained in the Appendix.) These are the sentences in the arithmetical component of S. In particular, we can talk about the truth or falsity of statements in the arithmetical component of the system without presupposing that such talk extends to arbitrary sentences in the language of the system. Thus, given that we can speak of the truth or falsity of statements in the arithmetical component of any formal system to which the incompleteness theorem applies, it follows, as noted, that the incompleteness theorem does indeed establish that there is, for these systems, a true arithmetical statement in the language of the system that is not provable in the system. Whether we can *tell*, for some particular undecidable statement A, which of A and not-A is true is another matter, and one that will be considered in Section 2.7 in connection with the discussion of Gödel's proof of the first incompleteness theorem.

Philosophical Misgivings about "True"

Of course such reflections on the truth or falsity of statements undecidable in a formal system S presuppose that we can in fact sensibly talk about arithmetical statements as being true or false. Very often in discussions of the incompleteness theorem it is regarded as unclear what might be meant by saying that an arithmetical statement which is undecidable, say in PA, is true. What, for example, are we to make of the reflection that the twin prime conjecture may be true, but undecidable in PA? In saying that the twin prime conjecture may be true, do we mean that it may be provable in some other theory, and if so which one? Do we mean that we may be able to somehow "perceive" the truth of the twin prime conjecture, without a formal proof? Or are we invoking some metaphysical concept of truth, say truth in the sense of correspondence with a mathematical reality?

The question is a natural one, since so many philosophical conundrums are traditionally formulated in terms of truth, and a question such as "What does it mean for an arithmetical statement to be true?" is almost automatically regarded as taking us into the realm of philosophical argument and speculation. Hence mathematicians, who usually avoid such argument and speculation like the plague, tend to put scare quotes around "true" or avoid the word altogether when discussing mathematics in a nonmathematical context.

In a mathematical context, on the other hand, mathematicians easily speak of truth: "If the generalized Riemann hypothesis is true...," "There are strong grounds for believing that Goldbach's conjecture is true...," "If the twin prime conjecture is true, there are infinitely many counterexamples. . . ." In such contexts, the assumption that an arithmetical statement is true is not an assumption about what can be proved in any formal system, or about what can be "seen to be true," and nor is it an assumption presupposing any dubious metaphysics. Rather, the assumption that Goldbach's conjecture is true is exactly equivalent to the assumption that every even number greater than 2 is the sum of two primes. Similarly, the assumption that the twin prime conjecture is true means no more and no less than the assumption that there are infinitely many primes p such that $p+2$ is also a prime, and so on. In other words "the twin prime conjecture is true" is simply another way of saying exactly what the twin prime conjecture says. It is a mathematical statement, not a statement about what can be known or proved, or about any relation between language and a mathematical reality.

Similarly, when we talk about arithmetical statements being true but undecidable in PA, there is no need to assume that we are introducing any problematic philosophical notions. That the twin prime conjecture may be true although undecidable in PA simply means that it may be the case that there are infinitely many primes p such that $p + 2$ is also a prime, even though this is undecidable in PA. To say that there are true statements of the form "the Diophantine equation $D(x_1, \ldots, x_n) = 0$ has no solution" that are undecidable in PA is to make a purely mathematical statement, not to introduce any philosophically problematic ideas about mathematical truth. (This particular purely mathematical statement is also a mathematical theorem, as will be explained in Chapter 3 in connection with the Matiyasevich-Robinson-Davis-Putnam theorem.)

Similar remarks apply to the observations made earlier regarding consistent systems and their solutions of problems. It was emphasized that the mere fact of a consistent system S proving, for example, that there are infinitely many twin primes by no means implies that the twin prime hypothesis is true. Here again it is often thought that such an observation involves dubious metaphysical ideas. But no metaphysics is involved, only ordinary mathematics. We know that there are consistent theories extending PA that prove false mathematical statements—we know this because this fact is itself a mathematical theorem—and so we have no mathematical basis for concluding that the twin prime conjecture is true, which is to say, that there are infinitely many twin primes, from the two premises "PA is consistent" and "PA proves the twin prime hypothesis."

Note that this use of "true" extends to the axioms of a theory. It is sometimes thought, when "true" is used in some philosophical sense, that the axioms of a theory cannot be described as true, since they constitute the starting point that determines what is meant by "true" in later discourse. All such philosophical ideas are irrelevant to the mathematical use of the word "true" explained above, which will be adhered to throughout the book when speaking of mathematical statements as true or false. For example, that the axiom "for every n, $n + 0 = n$" in PA is true means only that for every natural number n, $n + 0 = n$. In this case, indeed, we *know* that the axiom is true. Why and how we know this—whether by stipulation, by inspection, by intuition—is irrelevant to the meaning of the word "true" in this usage.

In the preceding comments, it has been emphasized that, for example, to say that Goldbach's conjecture is true is the same as saying that every even number greater than 2 is the sum of two primes, as "true" is used in this book. Thus, the use of "true" is in such cases just a convenience, freeing us from the need to repeat the formulation of the conjecture. But "true" is also used in the discussion in other ways, as when it is said that every theorem of PA is true. Here, we cannot eliminate the word "true" and replace the statement "Every theorem of PA is true" with a statement listing all of the infinitely many theorems of PA. It is still the case, however, that "Every theorem of PA is true" is a mathematical statement, not a statement about what can be proved or seen to be true, or a philosophical statement about mathematical reality. Alfred Tarski showed in the 1930s how to give a mathematical definition of truth on which a statement "A is true," for any given A, is mathematically equivalent to A itself. The formal details will not be presented in this book, but the discussion will rely on the possibility of giving such a mathematical definition of "true" in the case of arithmetic.

"True in S"

A fairly common use of "true" in popular formulations of the incompleteness theorem is illustrated by the following:

> Gödel proved that any set of axioms at least as rich as the axioms of arithmetic has statements which are true in that set of axioms, but cannot be proved by using that set of axioms.

Again, the Columbia Encyclopedia states that

> Kurt Gödel, in the 1930s, brought forth his incompleteness theorem, which demonstrates that an infinitude of propositions that are underivable from the axioms of a system nevertheless have the value of true within the system.

Clearly, "true in that set of axioms" or "true within the system" does not mean "provable from that set of axioms" or "provable in the system" here, but since there is no notion of "true in a set of axioms" in logic, the question arises what it does mean. One possible interpretation, on which the above formulations become intelligible but incorrect, will be considered in Chapter 7, in connection with the completeness theorem for first-order logic. However, it also appears that in many cases when such phrases as "true in the system" and "true within the set of axioms" are used, what the author means by this is that users of the system are somehow able to convince themselves of the truth of the statement, even though it is not formally derivable. In such a case it is a relevant observation that it may or may not be the case, depending on the system, that we know of any particular statement that it is true but unprovable in the system. In particular, we may or may not have any idea whether the undecidable sentence exhibited in Gödel's proof is true, as will be further commented on in Section 2.7.

A possibly related misunderstanding of the incompleteness theorem is found in formulations that speak of *theorems* as undecidable, as in "If the system is consistent, then there is a true but unprovable theorem in it." Every theorem of a system S is provable in S, since this is what "theorem of S" means. A similar odd usage is found in the claim that

> Gödel's incompleteness theorem states that in every mathematical system, there exists one axiom which can neither be proved nor disproved.

As has already been emphasized, every axiom of a system is trivially also provable in the system.

We should note that an unfortunate peripheral use of the phrase "unprovable theorems" exists in logic. The American logician (and professor of music) Harvey Friedman explains that

> An unprovable theorem is a mathematical result that can not be proved using the commonly accepted axioms for mathematics (Zermelo-Fraenkel plus the axiom of choice), but can be proved by using the higher infinities known as large cardinals. Large cardinal axioms have been the main proposal for new axioms originating with Gödel.

This unfortunate terminology does not of course imply that there are theorems of S that are unprovable in S, for any system S. Large cardinal axioms will be commented on in Chapter 8, which also gives a brief explanation of the direction of Friedman's work on the use of large cardinal axioms.

2.5 The First Incompleteness Theorem and Hilbert's *Non Ignorabimus*

We know from the incompleteness theorem that (assuming consistency) not even ZFC decides every arithmetical statement. Is it possible that, for example, the twin prime conjecture is in fact undecidable in ZFC? It's a logical possibility, but there is nothing to support it. No arithmetical conjecture or problem that has occurred to mathematicians in a mathematical context, that is, outside the special field of logic and the foundations or philosophy of mathematics, has ever been proved to be undecidable in ZFC.

So a mathematician who attacks a natural mathematical problem in the optimistic spirit of Hilbert's "non ignorabimus" is not obliged to feel at all worried by the possibility of the problem being unsolvable within ZFC.

Note the formulation "no arithmetical conjecture or problem." If we consider problems in *set theory* rather than arithmetic, the very first problem on Hilbert's list of 23 open mathematical problems, that of proving or disproving a set-theoretical conjecture known as Cantor's continuum hypothesis, *is* known to be unsolvable in ZFC (given that ZFC is consistent). That this problem is unsolvable in ZFC was proved using set-theoretical methods introduced by Gödel in 1938 and Paul Cohen in 1963, not with the help of the incompleteness theorem. Since ZFC is known to encompass all of the methods of "ordinary" mathematics, what this means is that in order to prove or disprove the continuum hypothesis, new mathematical axioms or principles of reasoning must be introduced. Since it is not part of ordinary mathematical activity to find and propose such axioms or principles, it is understandable that mathematicians in general tend to regard a problem that is known to be unsolvable in ZFC as no longer posing a mathematical problem. Similarly, if the twin prime hypothesis were to be proved to be undecidable in ZFC (which would be sensational in the extreme), the problem of its truth or falsity would thereby take on a different mathematical dimension.

Nevertheless, extensions of the axioms and rules of reasoning accepted in mathematics have occurred and are continually being explored, even if not as part of ordinary mathematical activity. So a Hilbert-style optimist may well take the view that the impossibility of formulating any one formal system within which every arithmetical problem is solvable does not exclude the possibility of every arithmetical problem being solvable in one or another of an indefinite series of further extensions of mathematics by new axioms or rules of reasoning. This indeed was the view of Gödel, who, as previously noted, proposed "large cardinal axioms" as a means of extending present-day mathematics.

2.6 The Second Incompleteness Theorem

We begin with an informal statement of the theorem:

Second incompleteness theorem (Gödel). *For any consistent formal system S within which a certain amount of elementary arithmetic can be carried out, the consistency of S cannot be proved in S itself.*

Let it first be noted that the "certain amount of arithmetic" is in this case not the same "certain amount" as in the first incompleteness theorem. This point will be commented on further in connection with the proof of the second incompleteness theorem.

The formulation of the second incompleteness theorem presupposes that the statement "S is consistent" can at least be *expressed* in the language of S, since otherwise the observation that the consistency of S cannot be proved in S would have no interesting content. (It is not an observation of any interest that a theory of arithmetic cannot prove that horses are four-legged.) It is perfectly possible to produce formal systems S whose language refers directly to sentences and proofs in formally defined languages, including the language of S. The second incompleteness theorem, however, only requires S to have an *arithmetical* component, as in the case of the first completeness theorem, and therefore presupposes some way of representing sentences and proofs as numbers and expressing statements about sentences and proofs as arithmetical statements about the corresponding numbers. This is known as the *arithmetization of syntax* and was first carried out by Gödel in his proof of the incompleteness theorem. A method of representing syntactical objects (such as sentences and proofs) as numbers is called a *Gödel numbering*. The most tedious part of a formal treatment of the incompleteness theorem consists in defining a Gödel numbering and in showing that "n is the Gödel number of a proof in S of the sentence with Gödel number m" can be defined in the language of arithmetic. It is required that the Gödel number of any sentence or sequence of sentences can be mechanically computed, and that computable properties of syntactic objects correspond to computable properties of Gödel numbers.

Details regarding Gödel numberings are just the kind of technicality that will be avoided in this book, although an example of a Gödel numbering will be given in Chapter 3. Mostly we will merely take for granted that a Gödel numbering can be introduced, so that the arithmetical component of a system S contains statements that we can interpret as being about sentences and proofs in formal systems, including the system S itself.

This notion of "can interpret" merits some further comments. Take as an example Con_S, a translation into the language of arithmetic of "S is consistent." That is, Con_S is an arithmetical statement obtained by introducing a Gödel numbering for the language of S, and expressing in the language of arithmetic "it is not the case that for some sentence A in the language of S, there is a proof in S both of A and of not-A." Con_S if written out as a sentence in the language of elementary arithmetic would be enormous, and we wouldn't be able to make sense of it if presented with it. The arithmetical sentence Con_S is not of any interest from an ordinary arithmetical point of view, but only in virtue of a conventional association between numbers and syntactic objects (such as sentences and sequences of sentences) whereby we know it to be true as an arithmetical sentence if and only if S is consistent. We *refer to* this sentence in reasoning that uses ordinary mathematical language; we do not *use* the sentence in formulating our reasoning. The same is true of other translations of statements about formal systems into the language of arithmetic using a Gödel numbering.

A similar situation is found, for example, in the use of binary data to represent sounds and pictures, say in computer games. Events in the game take place as a result of mathematical transformations of an enormous collection of bits (0 and 1) representing a certain situation in the game. These transformations, although describable in purely mathematical terms, make sense to us and are of interest only in virtue of a conventional association between bit patterns and certain sounds and images, and it is only in terms of sounds and images that we discuss these collections of bits.

An everyday example of an association between numbers and sequences of symbols is that used in basic arithmetic. We use sequences of digits such as "365" to denote numbers. The association between the sequence of digits "365" and the number 365 is a conventional one, resulting from a certain way of systematically interpreting sequences of symbols as numbers. We carry out various operations on these sequences of symbols which are describable in purely syntactic terms. For example, we use the operation of putting another "0" at the end of a string of digits. Although describable purely in terms of symbol manipulation, the interest of this operation lies in the fact that in virtue of our association of sequences of symbols with numbers, it corresponds to multiplication of a number by 10. Thus, we have in this case a "Gödel numbering in reverse," where statements about and operations on numbers can be expressed as statements about and operations on sequences of symbols.

The idea of statements about some kind of formally definable object—numbers, sets, sequences of symbols, bit patterns—being interpretable as statements about another kind of object is thus one that had been around for a long time before Gödel, and it is an idea used in many contexts other than formal logic. But Gödel put this idea to good use in the arithmetization of syntax, opening up a new approach in the study of formal systems.

Proving Con_S

That a formal system S is inconsistent means that there are two proofs in the system such that one proves A and the other proves not-A, for some sentence A. Since the property of being the Gödel number of a proof in S is required to be a computable one, it follows that "S is consistent" can be formulated as a Goldbach-like statement: it is not the case that there are numbers n and m such that n is the Gödel number of a proof in S of A and m is the Gödel number of a proof in S of not-A, for the same statement A. From this it follows that Con_S, if false, can be shown to be false by a computation, but if it is true it may or may not be provable using a given set of mathematical methods and principles. The second incompleteness theorem tells us that in fact Con_S, if true, cannot be shown to be true using only the methods and principles contained in the system S. But of course Con_S is provable in other formal systems, and it may or may not be provable in a system that we find mathematically justifiable.

There is a common misconception concerning the second incompleteness theorem, expressed for example in [Kadvany 89, p. 165], in the author's comments on supposed postmodernist implications of the incompleteness theorem:

> Gödel's Second Theorem implies that the consistency of *Principia* can be mathematically proven only by conjecturally assuming the consistency of *Principia* outright (which is what mathematicians implicitly do in practice), or by reducing the consistency of *Principia* to that of a stronger system, thereby beginning an infinite regress.

The second incompleteness theorem does not imply that the consistency of a system S can only be proved in a *stronger* system than S, if by a stronger system we mean a system that proves everything S proves and more besides. It only implies that the consistency of S cannot be proved in S itself. It would be strange indeed if the consistency of S could only be proved in a stronger system, since to say that S is consistent is only to say that S does not prove any contradiction—it may prove lots of false statements and yet be consistent. So a proof that S is consistent is not a proof that S is generally reliable as a source of arithmetical theorems, and there is no reason why a consistency proof for S has to presuppose the methods of reasoning of S itself. Thus, for example, the consistency of PA was proved by Gerhard Gentzen (1936) in a theory that extends PA in one respect and severely restricts it in other respects.

What is true is that a consistency proof for a consistent theory S can only be given in a theory S' which extends S with respect to the *Goldbach-like* statements provable in S. If S' proves the consistency of S, Con_S is a Goldbach-like theorem of S' that is not provable in S, while every Goldbach-like theorem of S is also provable in S'. (This is so because it is provable in S' that every Goldbach-like theorem of a consistent theory is true.)

The idea that the second incompleteness theorem leads to an infinite regress if we seek to prove the consistency of theories is often expressed. What is odd about this idea is that it has been well understood since antiquity that we cannot keep justifying our axioms and principles on the basis of other axioms and principles, on pain of infinite regress. Say we carry out a consistency proof for PA in a theory S. Why do we accept this proof? If we say that we need a consistency proof for S in order to accept the proof in S of the consistency of PA, then indeed we are on the way to an infinite regress. But we don't need Gödel to tell us that we cannot accept a proof in one formal system only on the basis of a proof in another formal system. At some point, we can only justify our axioms, and thereby our proofs, by informal means, whatever these may be—appeals to their intuitively clear and convincing character, to their usefulness or success in practice, to tradition, or to whatever else we may come up with in the way of justification.

Suppose the consistency of PA were in fact provable in PA itself. Would such a proof have any value as a proof that PA is in fact consistent? Not necessarily, for why should we accept this particular proof in PA? After all, if the consistency of PA is in doubt, the validity of a consistency proof given in PA is equally in doubt—at least as long as we haven't seen the proof. For there is a loophole: we might consider a consistency proof for PA carried out in PA to be conclusive because it only uses a small part of PA, one that we have no doubt is consistent even if we can muster doubts about the consistency of PA as a whole. This indeed was the general approach of David Hilbert.

The Second Incompleteness Theorem and Hilbert's Program

The second incompleteness theorem had profound consequences for the ideas concerning the foundations of mathematics put forward by David

Hilbert, who had the goal of proving the consistency of mathematics by *finitistic* reasoning. What he wanted to do was to formulate formal systems within which all of ordinary mathematics could be carried out, including the mathematics of infinite sets, and to prove the consistency of these systems using only the most basic and concrete mathematical reasoning. In this sense, Hilbert intended his consistency proof to be conclusive: the consistency of, say, ZFC is proved using only reasoning of a kind that we cannot do without in science or mathematics.

Although Hilbert did not formally specify what methods were allowed in finitistic reasoning, it seems clear that the methods Hilbert had in mind can be formalized in systems of arithmetic such as PA. And if PA cannot prove its own consistency, it follows that not even the consistency of elementary arithmetic can be proved using finitistic reasoning, so that Hilbert's program cannot be carried out (and that the finitistic consistency proof for arithmetic that Hilbert thought at the time had already been achieved, by his student and collaborator Wilhelm Ackermann, must be incorrect, as indeed it turned out to be).

It is often said that the incompleteness theorem demolished Hilbert's program, but this was not the view of Gödel himself. Rather, it showed that the means by which acceptable consistency proofs could be carried out had to be extended. Gödel's own "Dialectica interpretation," which he developed in the early 1940s and which was published in the journal *Dialectica* in 1958, gave one way of extending the notion of finististic proof.

Hilbert's program was based on a particular set of ideas about the meaning and contents of mathematics and what is meant by a justification of mathematics. If we do not share Hilbert's general ideas, a consistency proof is by itself quite insufficient to justify a mathematical theory, since, for example, provability in a consistent theory of the twin prime conjecture is no guarantee that there are infinitely many twin primes. The nonmathematical conclusions drawn from the second incompleteness theorems in recent decades are rarely explicitly based on Hilbert's views, but the skepticism they tend to express is clearly influenced by similar ideas. Such conclusions will be commented on at some length in Chapter 5.

2.7 Proving the Incompleteness Theorem

It is sometimes thought that the first incompleteness theorem is not a mathematical theorem in the ordinary sense. For example:

> The theorem is written in a formal mathematical system, but can only be proved using nonformal mathematical reasoning. There is no way to formalize the proof; it can only be stated in a natural language like English.
>
> Another difficulty is whether Gödel's proof is actually a proof. To prove incompleteness, we have to interpret the formula and have to understand that what it says is true. That is, the result is not achieved by formal reasoning, but by some meta-reasoning done from outside the system. Hence it is not a "formal" proof. It takes insight to see the truth of the formula.

This idea that Gödel's theorem does not have an ordinary mathematical proof seems to be based on a specific misunderstanding of Gödel's original proof of the first incompleteness theorem: the mistaken belief that Gödel's proof, which shows a certain arithmetical statement G depending on S to be unprovable in S if S is consistent, also shows G to be *true*. This would indeed make Gödel's proof a remarkable one, but in fact the proof does not show anything of the kind. All that Gödel's proof shows is the implication "if S is consistent, G is true." If we can prove that S is consistent, which we can sometimes do, we can also prove that G is true. If we have no idea whether or not S is consistent, Gödel's proof still goes through, but we have no idea whether or not G is true. In either case, the question of the truth or falsity of G cannot be decided on the basis only of Gödel's proof.

In fact there is nothing any more informal or intuitive about the proof of Gödel's theorem than there is about mathematical proofs in general. The incompleteness theorem has formal proofs in fairly weak mathematical theories—PA is more than sufficient. Thus, applied to PA and stronger theories, Gödel's proof does not establish the truth of any mathematical statement which is not provable in the theory itself.

Gödel's Proof

Gödel's original proof of the incompleteness theorem was not formulated as a theorem about formal systems in general, for the reason that the general theory of computability, and therewith the general concept of a formal system, had yet to be formulated in 1930. Instead, Gödel proved the theorem for a particular formal system which he called P. He listed the properties of P used in the proof, and noted that these properties were shared by a wide class of formal systems. In Part II of the paper, which never appeared, he intended to formulate a general version of the theorem. During the 1930s it became clear, with the introduction of the general concept of computability, that essentially the same proof did indeed apply to all formal systems satifying the conditions listed by Gödel, and also that those conditions were satisfied by all formal systems in which a certain modest amount of arithmetic can be carried out.

Gödel's proof of the incompleteness theorem introduced a technique that has since been used extensively in logic, that of defining *provable fixpoints* for various properties of arithmetical sentences. This technique presupposes the arithmetization of syntax. Suppose we have defined in arithmetic some property P of the Gödel numbers of sentences in the language of the system S. For example, P could be the property of being the Gödel number of a Goldbach-like sentence in the language of S, or the property of being the Gödel number of an axiom of S, or the property of being the Gödel number of a theorem of S. We need to assume that the property P is itself definable in the language of arithmetic, as is the case with the properties mentioned. By a provable fixpoint for the property P is meant an arithmetical sentence A such that S, or a weaker system than S, proves

$$A \text{ if and only if } m \text{ has the property } P$$

for a particular number m, and this number m is in fact the Gödel number of A itself. Thus, it is provable in S that A holds if and only if its Gödel number has the property P.

That there is a provable fixpoint for every property P definable in the language of arithmetic is far from obvious. Gödel established the existence of such fixpoints by translating a statement that "says of itself that it has property P" into arithmetic. For this, he used a construction which in ordinary (or not so ordinary) language can be formulated as

> The result of substituting the quotation of "The result of substituting the quotation of x for 'x' in x has property P." for 'x' in "The result of substituting the quotation of x for 'x' in x has property P." has property P.

This oddly worded sentence says that the result of carrying out a certain specific substitution operation has property P. If we carry out the operation specified, we find that it results in precisely the sentence itself. The sentence therefore "says of itself that it has property P," in the sense that it says that a sentence satisfying a certain description has property P, and the sentence itself is the one and only sentence satisfying that description. By showing that the substitution operation can be defined in arithmetical terms as an operation on Gödel numbers, Gödel obtained a version A in the language of arithmetic of the statement above, provably equivalent in PA to "m has property P," where m is the Gödel number of A.

Other ways of formulating self-referential sentences using syntactic operations have also been formulated later. For example, the American logician and philosopher W. V. O. Quine came up with a method known as "quining":

> "yields a sentence with property P when appended to its own quotation." yields a sentence with property P when appended to its own quotation.

The general fixpoint construction is widely used in logic to prove various results. Gödel used it to prove his first incompleteness theorem, by applying it to the property of not being a theorem of S. By a *Gödel sentence* for S is meant a sentence G obtained through the general fixpoint construction, such that S proves

> G if and only if n is not the Gödel number of a theorem of S,

where n is the Gödel number of G itself.

Let us first observe that G can be formulated as a Goldbach-like statement. It is equivalent to the statement that no number p is the Gödel number of a proof of G in S, and the property of being such a number p is a computable one, given the general requirements on formal systems and Gödel numberings.

The reasoning in Gödel's proof is now as follows. First, if G is in fact a theorem of S, then it is provable in S that G is a theorem of S (that is, that n is the Gödel number of a theorem of S). The reason for this is that being a theorem of S is a property that can be verified by exhibiting a proof in S, and since being a proof in S is required to be a computable property of sequences of sentences, the verification can be carried out within S. So if G is a theorem of S, this is provable in S, but since G is a provable fixpoint of the property of *not* being a theorem of S, the negation of G is then also provable in S, so S is inconsistent.

Thus, if S is consistent, G is not provable in S. Can we also conclude that G is not *disprovable* in S, on the assumption that S is consistent? No, for by the second incompleteness theorem, a consistent theory S may prove its own inconsistency, and thereby prove that *every* statement is provable in S (and in particular, prove G false, since G is equivalent to the statement that G is not provable in S). But we can observe that if S is consistent, G is true (because not provable in S), and since G is also a Goldbach-like sentence, it follows that not-G is not provable in S provided we assume that S is Σ-sound, that is, does not disprove any true Goldbach-like sentences. This yields a result that is a bit stronger than Gödel's original version of the first incompleteness theorem, since Gödel's original assumption that S is ω-consistent implies, but is stronger than, the assumption that S does not disprove any true Goldbach-like sentences.

Rosser Sentences

Rosser, in order to strengthen the formulation of the incompleteness theorem, introduced a *Rosser sentence* R, which is constructed as a fixpoint of a more complicated property than that of not being provable in S. Specifically, PA proves

> R if and only if for every n, if n is the Gödel number of a proof of R, then there is an $m < n$ such that m is the Gödel number of a proof of not-R.

It is left to the interested reader to verify that R is also a Goldbach-like statement, and furthermore is undecidable in S if S is consistent. Nothing essential will be lost by just taking this fact on faith in the following.

Tarski's Theorem

The fixpoint construction can be used to show that there is no way of defining in the language of arithmetic the property of being the Gödel number of a *true* arithmetical sentence. For if this property could be defined in arithmetic, a fixpoint A for the property of *not* being the Gödel number of a true arithmetical sentence would have the property that A is a true arithmetical sentence if and only if its Gödel number is not the Gödel number of a true arithmetical sentence. In other words, a form of the ancient paradox of the Liar would be a consequence: we would have constructed an arithmetical statement A that is true if and only if it is not true. The result that the property of being a true arithmetical sentence cannot be defined in the language of arithmetic is usually called Tarski's theorem, but in fact Gödel obtained this result as a step along the way in discovering the incompleteness theorem.

Although the property of being a true arithmetical sentence thus cannot be defined in arithmetic, the property of being a true Goldbach-like sentence *can* be so defined, and similarly for other restricted categories of arithmetical sentence. Thus, for example, we can define an arithmetical sentence A that is true if and only if it is not a true Goldbach-like sentence. This does not result in any paradox, since A is in fact not a Goldbach-like sentence at all, and is therefore true.

Gödel himself observed that his proof of the incompleteness theorem is related to the paradox of the Liar in the following sense. The Liar sentence is a sentence L that "says of itself that it is not true." Thus, given our ordinary use of "true" it seems to follow that L is true if and only if it is not true, a logical contradiction. Gödel's proof uses a corresponding arithmetical sentence in which "true" is replaced by "provable in S." Whereas the paradox of the Liar has given rise to endless debates over the meaning of "true" and the question what is required for a sentence to express a meaningful assertion, the Gödel sentence is an arithmetical sentence, one that is as unproblematically meaningful as other arithmetical sentences of a similar logical form (Goldbach-like statements).

Gödel's original proof is by no means the only proof of the first incompleteness theorem, and some other proofs will be briefly commented on at a later point. But first some remarks about self-reference, which the reader may well prefer to skip.

Self-Reference

Gödel's proof of the first incompleteness theorem, and Rosser's strengthened version, both make essential use of the fixpoint construction. It is sometimes claimed that it is misleading to say that sentences obtained through the fixpoint construction are "self-referential," since (it is held) all that can be said is that the equivalences

$$A \text{ if and only if } m \text{ has the property } P$$

are provable for these sentences A, where m is the Gödel number of A itself. In other words, sentences like G and R are just fixpoints for certain properties, they are not "self-referential."

This is not correct, however. Gödel's *proof* does indeed only use the fact that the Gödel sentence G is a fixpoint for the property of not being provable in S. But the sentences constructed in the proof that every arithmetical property P has a provable fixpoint are self-referential in a stronger sense: they are sentences A of the form

There is an m such that m has property P and property Q

where it is provable in PA that the only number that has property P is the Gödel number of the sentence A itself. It is in this sense that the sentence A "says of itself that it has property Q." There are provable fixpoints other than Gödel sentences for the property of not being provable in S, for example (by the second incompleteness theorem) the sentence Con_S formalizing "S is consistent." Since it is provable in S that Con_S is true if and only if it is not provable in S, Con_S is in fact a fixpoint for the property of not being provable in S. But Con_S is not in any apparent sense self-referential, and nobody would want to say that Con_S "says of itself that it not provable in S."

This characterization of self-referential arithmetical statements doesn't tell the whole story. Let us consider a couple of varieties of ordinary self-reference, in the sense of speakers referring to themselves, rather than sentences referring to themselves. Suppose John says "John loves you." This may or may not be a case of self-reference, depending on the speaker's intention. He may be referring to himself in the third person the way George Costanza does in *Seinfeld*, when observing that "George is getting angry," or he may be referring to another person named John. On the other hand, if John says "Your husband loves you," and is addressing his wife, his statement is self-referential in a sense that is independent of his intentions. Even if he is an amnesiac and makes the statement on the basis of what he has learned about his wife's husband, it is still self-referential in the sense that John is the one and only husband of the person addressed, and therefore the statement is true if and only if John loves his wife.

In the case of self-referential statements in the language of arithmetic, there is of course no question of the sentence itself intending anything, but we can locate two similar aspects of self-reference. The use of substitution operations or something similar corresponds to the second example of self-reference above, while the first example corresponds to the conventional choice of Gödel numbering. Suppose we formulate a statement A

0 has property Q

and then introduce a Gödel numbering in which the Gödel number of A is 0. There is no problem about introducing such a Gödel numbering, since a Gödel numbering is purely a matter of convention, as long as computable syntactic properties correspond to computable properties of Gödel numbers. The statement A is then, given this Gödel numbering, trivially a provable fixpoint for the property Q. Indeed, in the sense explained, A does "say of itself that it has the property Q." It has the form "there is an m such that m has property P and property Q," where the Gödel number of A itself is the only number with property P (namely that of being identical with 0).

Something seems to be missing in this example of self-reference. Why does this approach not yield a painless version of Gödel's proof? The reason is that we first choose the property Q and then the Gödel numbering, depending on Q. We can't apply this construction to obtain a provable fixpoint for the property of not being a theorem of A, since the latter property itself depends on a choice of Gödel numbering. The trivial construction does yield a statement which "says of itself that it has the property Q," but the property Q is in this case just a random and pointless property of the Gödel number of the sentence. A statement that is self-referential in the sense that G and R are self-referential not only has the form "There is an m such that m has property P and property Q," where the Gödel number of the statement itself is the only number with property P. In addition, P and Q are translations into arithmetic, using the same Gödel numbering, of syntactical properties of statements.

Proofs of the First Incompleteness Theorem Using the Theory of Computability

Gödel's proof of the first incompleteness theorem and Rosser's strengthened version have given many the impression that the theorem can only be proved by constructing self-referential statements in the language of S, or even that only strange self-referential statements are known to be undecidable in elementary arithmetic.

To counteract such impressions, we need only introduce a different kind of proof of the first incompleteness theorem. One argument is as follows. By the Matiyasevich-Robinson-Davis-Putnam theorem, which has been referred to a couple of times already, there is no algorithm that given any Diophantine equation $D(x_1, \ldots, x_n) = 0$ will decide whether or not the equation has a solution. But then there can be no theory that correctly decides every statement of the form "the Diophantine equation $D(x_1, \ldots, x_n) = 0$ has at least one solution." For given such a theory S, we could decide whether or not a given equation $D(x_1, \ldots, x_n) = 0$ has a solution by systematically searching for a proof in S of either the statement "the equation $D(x_1, \ldots, x_n) = 0$ has at least one solution" or its negation. Thus, unless S proves some false sentence of the form "the equation $D(x_1, \ldots, x_n) = 0$ has at least one solution," there must be sentences of this form that are undecidable in S.

Note two features of this argument. First, like Gödel's original proof, it only shows incompleteness on the assumption that S does not incorrectly disprove any true Goldbach-like statements. But second, it does not, as it stands, exhibit any particular statement undecidable in S, unlike Gödel's proof. On the other hand, we don't even need to introduce any arithmetization of syntax in order to conclude the incompleteness of S by this argument, since the arithmetical incompleteness of S is located among ordinary arithmetical statements about Diophantine equations. By introducing arithmetization, the argument can be refined to exhibit a particular undecidable sentence, as in Gödel's original proof.

This proof of the first incompleteness theorem, along with the Matiyasevich-Robinson-Davis-Putnam theorem, will be explained in greater detail in Chapter 3, which is devoted to the theory of computability and its connections with the incompleteness theorem. A third approach to proving the first incompleteness theorem, using the concept of Kolmogorov complexity, will be presented in Chapter 8.

A reader who is disinclined to digest the reasoning in these other proofs of the first incompleteness theorem should just keep in mind the essential point that incompleteness is not restricted to arithmetical translations of strange self-referential statements. Instead, we know that systems S to which Gödel's theorem applies have undecidable statements of the form "the Diophantine equation $D(x_1, \ldots, x_n) = 0$ has no solution." What we do not know is whether any statements of this form of interest to mathematicians are undecidable in PA or ZFC.

Weaker Variants of the First Incompleteness Theorem

There are still other proofs of the incompleteness of PA and of the arithmetical component of other theories. Thus, [Putnam 00] presents one such proof, due to Saul Kripke. Unlike the proofs already described, this proof of Kripke's neither makes use of self-referential statements nor draws on the theory of computability. (Instead it makes use of the existence of *nonstandard models* of PA, explained in Section 7.3.) However, the conclusion proved is weaker than that in Gödel's theorem with regard to the logical complexity of the statement shown to be undecidable. It is not a Goldbach-like statement, and in particular the argument does not establish the truth of the second incompleteness theorem.

Proving the Second Incompleteness Theorem

Gödel's proof of the second incompleteness theorem in his 1931 paper consisted mostly of handwaving—in other words, sketching an argument without carrying it out in detail. The argument was simple: the proof of the first incompleteness theorem established that if the system P is consistent, G is not provable in P, and therefore true. If we examine this argument, we see that it only uses mathematical reasoning of a kind that can be carried out within P, and therefore it follows that the implication "if P is consistent then G" is provable in P. But then, if P is consistent, it follows that "P is consistent" is not provable in P, since G is not provable in P.

We can strengthen this conclusion: not only "if P is consistent then G" is provable in P, but also "if G then P is consistent." For G is equivalent in P to "G is not a theorem of P," and every theory in which some statement is not provable is consistent, a fact which is also provable in P. Thus, G and "P is consistent" are in fact equivalent in P.

As in the case of the first incompleteness theorem, it was clear from Gödel's presentation that the argument extended to a wide class of formal systems incorporating "a certain amount of arithmetic," although the "certain amount" is not the same "certain amount" as in the first incompleteness theorem. Gödel's proof of the second incompleteness theorem for a formal system S depends on the proof of "if S is consistent, G is unprovable in S" being formalizable in S itself. A formal system needs to incorporate a larger amount of arithmetic in order for the proof of the first incompleteness theorem to be *formalizable* in the system than it needs for the proof of the first incompleteness theorem to *apply* to the system. Thus, we can specify formal systems such as Robinson Arithmetic, presented in the Appendix, to which the first incompleteness theorem applies, but not the second. Note that this does not mean that Robinson arithmetic proves its own consistency, but only that we cannot use the second incompleteness theorem, as formulated here, to show that it does not prove its own consistency. (Formulations of arithmetical theories that do prove their own consistency exist in the logical literature—these theories are in one way or another very weak, for example, in not assuming that every natural number has a successor, but can nevertheless incorporate a great deal of nontrivial arithmetical reasoning.)

Gödel's argument was in fact quite convincing to his readers, and he never got around to presenting his proof of the second incompleteness theorem in full formal detail. This was instead done in the 1939 two-volume work *Grundlagen der Mathematik* (Foundations of mathematics) by David Hilbert and Paul Bernays. Gödel's original proof remains the most common and the most accessible proof of the second incompleteness theorem, although other proofs exist for particular theories.

As in the case of the first incompleteness theorem, it is sometimes thought that the proof of the second incompleteness theorem is not an ordinary mathematical proof. Thus, Kadvany comments [Kadvany 89, p. 178] that "The historical process shows that the Second Theorem is not a theorem in the ordinary sense." This is not an idea based on a simple misunderstanding of the theorem, as in the case of similar comments about the first incompleteness theorem. Instead, what Kadvany has in mind is the fact that the second incompleteness theorem requires us to decide on a formalization of "S is consistent" in the language of arithmetic. When we consider *arbitrary* formal systems this becomes a somewhat delicate matter, and it only became clear around 1960, through the work of the American philosopher and logician Solomon Feferman, how to formulate the second incompleteness theorem in the most general case. However, these subtleties do not affect the formulation of the theorem applied to PA or ZFC or any other theory that we actually study in logic. In these cases, there is essentially only one way of expressing "S is consistent" as an arithmetical statement, namely the obvious and natural way. Kadvany remarks in this context that "What is natural is a matter of historical choice made against a background of mathematical tradition adapting to a radical new set of ideas," (p. 176) which is misleading. In the general case (arbitrary theories S) there simply *is* no natural choice, and no historical choice has been made. In the case of theories like PA or ZFC, the choice of arithmetical formulation of "S is consistent" is natural, given a particular choice of Gödel numbering, in the same sense as the choice of an arithmetical formulation of, for example, the fundamental theorem of arithmetic ("every natural number has a unique decomposition as a product of primes") is natural, given a representation of finite sequences of numbers as numbers. No difficult or arbitrary choices are involved, beyond the choice of a Gödel numbering. Furthermore, although the choice of a Gödel numbering has its arbitrary aspects, we can easily isolate the *standard* Gödel numberings, which yield equivalent Gödel sentences and equivalent formalizations of "S is consistent" for these theories.

As these comments probably suggest to the reader, this particular issue raised by Kadvany is somewhat technical, both logically and philosophically, and it will not be pursued in the remainder of this book. A reader interested in the subject will find an extended treatment in [Franzén 04].

2.8 A "Postmodern Condition"?

Provability or decidability, in the context of the incompleteness theorem, means provability or decidability in some formal system or another. There is no concept in logic of a statement A being provable or disprovable or undecidable in any absolute sense, but only of A being provable or disprovable or undecidable in a formal system S. Any statement A undecidable in S will be decidable in other systems. In particular, if A is undecidable in a consistent system S, A is provable in the consistent system $S + A$ obtained by adding A as a new axiom to S, and disprovable in the consistent system $S+$ not-A obtained by adding the negation of A as an axiom. Of course this does not tell us whether A is decidable in the more interesting sense of being provable or disprovable by reasoning that mathematicians would consider conclusive, or whether A is decidable in some particular theory of interest such as PA or ZFC.

It is sometimes thought that the fact of some particular statement being undecidable in some particular formal system implies that mathematics "branches off" into a version in which the statement is provable and one in which it is disprovable. An analogy is often suggested with the logical independence of the parallel postulate from the other axioms of Euclidean geometry, which led to axiomatizations of non-Euclidean geometries in which the parallel axiom does not hold. This idea is taken up in combination with a general view of Gödel's theorem as congenial to postmodernism in [Kadvany 89, p. 162]:

> The simplest observation of how Gödel's Theorems create a postmodern condition begins with the First Incompleteness Theorem. This theorem says, in effect, that a consistent axiomatic system strong enough to prove some weak theorems from elementary number theory, requiring only the operations of addition and multiplication, but not either operation separately,

> will be *incomplete*: there will always be mathematical sentences formulated in the syntax of the system under consideration that are neither provable nor refutable in the system, and these sentences are said to be *undecidable* with respect to the system. Since an undecidable proposition and its negation are each separately consistent with the base system, one can extend the old system to two mutually incompatible new ones by adding on the undecidable sentence or its negation as a new axiom. The classical example of this procedure is the generation of non-Euclidean geometries by adding the negation of the parallel postulate to the axioms of elementary geometry without the parallel postulate. The new systems so constructed also have new undecidable sentences, different from the originals, and the process of constructing new undecidable sentences and then new systems incorporating them or their negations goes on *ad infinitum*, like a branching tree which never ends.

Let us take a closer look at the suggested analogy. Axiomatic theories of geometry obtained by replacing the parallel postulate of Euclidean geometry with an incompatible axiom are, like Euclidean geometry itself, both mathematically interesting and useful in applications. In particular, Euclidean plane geometry is applicable to smaller parts of the Earth's surface, while the geometry of the Earth's surface in the large is non-Euclidean. It should be noted that a useful alternative to Euclidean geometry is not obtained simply by replacing the parallel postulate with its negation, but by introducing a more specific axiom incompatible with the parallel postulate. In the case of elliptic geometry, the geometry of the surface of a sphere (and thus approximately the geometry of the surface of the Earth), the parallel postulate is replaced by the axiom that there is *no* line parallel to a given line containing a point outside that line.

We can easily convince ourselves that the analogy with the parallel postulate does not apply in *every* case of undecidability. For example, PA has an axiom stating that for every n, $n + 0 = n$. This axiom is not provable from the remaining axioms of PA, so we get a consistent theory by changing the axiom to "it is not the case that for every n, $n + 0 = n$." By Gödel's completeness theorem for first-order logic (explained in Chapter 7), the resulting theory has a *model*, that is, it is possible to specify a mathematical structure in which there are e such that $e + 0$ is not equal to e, but which otherwise satisfies the axioms of PA. Unlike the case of non-Euclidean geometry, such a mathematical structure is not of any particular interest, either mathematically or from the point of view of applications. And of course the existence of such structures—there are very many of them—in no way affects the observation that $n + 0$ *is* equal to n for every natural number n.

We can also create an infinite tree of consistent variants of PA by omitting the axiom that $n + 0 = n$ and adding $0 + 0 = 0$ or its negation to PA, then to each of the two resulting theories add either $1 + 0 = 1$ or its negation, and so on. Thus we have one theory postulating that $0 + 0$ equals 0 and $1 + 0$ does not equal 1, one theory postulating that $0 + 0$ does not equal 0 but $1 + 0$ equals 1, and so on. No "postmodern condition" is created by this, but only an infinite tree of very uninteresting theories. (In Chapter 7 it is explained how we know that every theory in the tree is consistent.)

Thus, if we think that the incompleteness theorem opens up a spectrum of possible varieties of arithmetic, perhaps in the spirit of postmodernism, this must be because of some decisive difference between the trivial examples given and the incompleteness exhibited by Gödel's theorem. What decisive difference might this be?

The infinite tree of theories described is uninteresting in part because all statements $n + 0 = n$ are *true* of the natural numbers, and there is no apparent point in introducing theories that deny some or all of these statements. The resulting theories have no interesting application and are mathematically useless. Now consider the "branching" or "bifurcation" of PA based on Gödel's proof of the first incompleteness theorem. The undecidable statement G produced in Gödel's proof is equivalent in PA to "PA is consistent," and we do indeed get two consistent extensions of PA by adding either "PA is consistent" or "PA is not consistent" as a new axiom to PA. Similarly, for the resulting theory T, we get two consistent extensions by adding either "T is consistent" or "T is inconsistent" as a new axiom, and so on. But as in the trivial example, since all of the statements "PA is consistent," "T is consistent," and so on are known to be *true*, there is no obvious reason why we should take any interest in theories obtained by adding "PA is inconsistent" or "T is inconsistent." There is only one branch in this infinite tree that is of any immediately apparent interest, the sequence of theories PA, PA_1, PA_2,... obtained by adding axioms "PA is consistent" to get PA_1, "PA_1 is consistent" to get PA_2, and so on. The interest of this branch of the tree lies in the fact that it yields an extension of PA by an infinite number of true statements not provable in PA.

A comment is in order regarding a somewhat technical topic that will be considered further in Chapter 7. A mathematical structure in which the axioms of PA + not-"PA is consistent" are true is of a kind known as a *nonstandard* model of arithmetic: it contains, in addition to the "standard" natural number 0, 1, 2,... also "infinite elements," which are not natural numbers. Such structures do have mathematical interest, but this does not imply that we need to introduce theories with false arithmetical axioms in order to obtain such structures.

How do we know that all of the consistency statements generated in the way described are true? It is sometimes thought that we must here invoke some form of intuitive insight of a dubious nature and claim that we can "see" that the theories are consistent. In fact, it is *provable* in perfectly ordinary mathematics that these consistency statements are true, as a consequence of PA being a *sound* theory, or in other words, a theory all of whose axioms are true. Thus, there is no special "seeing" involved, other than the kind of "seeing" the truth of arithmetical statements that is involved in mathematical proofs of arithmetical statements in general. Now, it is of course open to anybody to put this proof into question, and take a skeptical view of the theorem that PA, PA_1, and the other theories are consistent. In this, the theorem is no different from other mathematical theorems of a comparable degree of abstraction. For example, we might take a skeptical view of Wiles' proof of Fermat's last theorem, and argue that we don't really know that the theorem is true on the basis of that proof. But such skepticism, whether or not it has anything to recommend it, is no more a matter of concern in connection with consistency proofs than it is in other mathematical contexts.

Suppose we replace PA in the construction described with ZFC. It is then no longer the case that the consistency of all the theories in the tree is provable in perfectly ordinary mathematics. (The only known proofs of the consistency of ZFC use set-theoretical axioms that are not part of ordinary mathematics.) But we still don't get any "branching" of arithmetic into different directions. The theory obtained by adding "ZFC is inconsistent" to ZFC is indeed consistent, given that ZFC is consistent, but it is of no apparent mathematical interest and has no application.

We can look for other examples, where we simply don't know whether the undecidable statements that are introduced are true or not. Using the

Matiyasevich-Robinson-Davis-Putnam theorem (explained in Chapter 3) and results in set theory, we can in fact produce a set of statements of the form "the Diophantine equation $D(x_1, \ldots, x_n) = 0$ has no solution" such that an infinite tree of the kind described can be constructed, starting from PA, for which we just don't know, for any of the statements involved, whether it is true or not. Again, this does not produce any "postmodern condition" in mathematics, since the theories obtained have no theoretical interest and no application. We just have a bunch of statements of the form indicated that we know to be unprovable in current mathematics, some of which may possibly turn out to be of some interest, and may or may not eventually be proved true or false.

In brief, the point argued is that the incompleteness theorem has not led to any situation in which mathematics (and specifically arithmetic) branches off into an infinity of incompatible directions, and there is no reason why it should. It is not commonly thought that our thinking about the place of humanity in the universe branches off into two different directions, one in which it is postulated that there is extra-terrestrial life within a thousand light years of our planet and one that postulates that there is not, and that these branches are further subdivided according to further postulates (consistent with our current knowledge) that might be made. The mere fact of incompleteness does not bring about any "branching off" in mathematics either.

Kadvany, in the article quoted, comments not only on undecidable statements resulting from the proof of the incompleteness theorem, but on statements known to be undecidable in *set theory* on the basis of the work by Gödel and Cohen and others. The continuum hypothesis has been mentioned earlier as a prime example of a statement that is not only known to be undecidable in ZFC, but which is not decided by any set-theoretic principle that any significant number of mathematicians regard as evident. Although this has not led to any branching off of set theory in different directions, comparable to the development of non-Euclidean geometries, there is greater scope than in the case of arithmetic for an argument that such a branching might take place in the future. However, this incompleteness of set theory does not arise from the incompleteness phenomenon revealed by Gödel's theorem and will not be considered further in this book, except for some incidental remarks in Chapter 8.

The ideas touched on here will be discussed further in Chapter 5, in connection with the skeptical views often associated with the second incompleteness theorem.

2.9 Mind vs. Computer

An argument usually associated with the British philosopher J. R. Lucas fastens onto the fact that we know that the Gödel sentence of (for example) PA is true, since we know that the theory is consistent, and concludes that the human mind surpasses any machine ([Lucas 61]):

> However complicated a machine we construct, it will, if it is a machine, correspond to a formal system, which in turn will be liable to the Gödel procedure for finding a formula unprovable in that system. This formula the machine will be unable to produce as true, although a mind can see that it is true.

This argument is invalid because based on the mistaken idea that "Gödel's theorem states that in any consistent system which is strong enough to produce simple arithmetic there are formulas which cannot be proved in the system, but which we can see to be true." The theorem states no such thing. As has been emphasized, in general we simply have no idea whether or not the Gödel sentence of a system is true, even in those cases when it is in fact true. What we know is that the Gödel sentence is true if and only if the system is consistent, and this much is provable in the system itself. When we know that the system is consistent, we also know that its Gödel sentence is true, but in general we don't know whether or not a formal system is consistent. If the human mind did have the ability to determine the consistency of any consistent formal system, this would certainly mean that the human mind surpasses any computer, but there is no reason whatever to believe this to be the case.

Given that we cannot conclude from the incompleteness theorem that the human mind surpasses any computer (or equivalently, any formal system) as far as arithmetic is concerned, we might attempt to draw the weaker conclusion that "no machine will be an adequate model of the mind" in the sense that no machine, although it may perhaps surpass the human mind in arithmetic, can ever be *exactly* equivalent to the human mind as far as arithmetical ability is concerned. But this too fails to follow from the incompleteness theorem. Let us assume (a large assumption) that there is such a thing as "human arithmetical ability," and go on to suppose that a particular formal system S exactly embodies that ability. If we know S to be consistent, we will indeed have a conflict with the incompleteness theorem. But again what is missing is an argument for *why* we should know S to be consistent. Lucas [Lucas 61] here introduces the irrelevant reflection:

"The best we can say is that S is consistent if we are." This is irrelevant because even if we know ourselves to be consistent, there is no reason why we should conclude from this that S is consistent, unless we already *know* that S codifies human arithmetical ability—and why should we know this? Thus, Gödel himself commented that nothing rules out the existence of a formal system S that exactly codifies human arithmetical ability, although we could not recognize the axioms of S as evidently true.

So let us weaken the conclusion further. What does follow from the incompleteness theorem is that we cannot actually specify any formal system S such that we *know* that S embodies those and only those arithmetical truths that we can convince ourselves are true. For given any system S for which we know that all its arithmetical theorems are true, we can produce an arithmetical statement—an arithmetization of "S is consistent"—which we also know to be true and which is not a theorem of S. Thus, we cannot specify any one formal system which exhausts all of our arithmetical knowledge.

This last argument, which was put forward by Gödel himself, seems to point to the *inexhaustibility* of our mathematical knowledge, which is one of the most striking consequences of the incompleteness theorem. It may be held that this in itself is an indication that human mathematical knowledge cannot be described in terms of mechanisms and computations. This argument will be considered in Chapter 6.

2.10 Some Later Developments

As noted earlier, nothing in the proof of the incompleteness theorem tells us whether the statements undecidable in PA or ZFC include such natural mathematical statements as Goldbach's conjecture. It would be of very great interest if any of the classical problems of number theory could be shown to be unsolvable in PA, but the methods and results described in this chapter give no clue as to whether this is the case.

In recent decades, there has been considerable work seeking to establish, as a consequence of the incompleteness theorem, the undecidability in PA or in other theories of arithmetical statements which, although not expressing conjectures that have occurred to mathematicians in nonlogical contexts, are still closer to "ordinary mathematics" than Gödel statements and consistency statements. The first result of this kind was the *Paris-Harrington theorem* (1976), establishing the unprovability in PA of the arithmetization of a certain combinatorial principle. This principle, although unprovable in PA, is in fact equivalent in PA to the statement that PA disproves no false Goldbach-like statement, and so is provable using ordinary mathematical reasoning. But we know that ZFC is also incomplete in its arithmetical component. A way of extending the arithmetical component of ZFC was suggested by Gödel in the 1940s and has been extensively studied since. This approach consists in extending ZFC with *axioms of infinity*, strong set-theoretical principles that imply arithmetical statements not provable in ZFC.

Another line of development, dating from the 1960s, relates incompleteness to the theory of what is known as Kolmogorov complexity. This development is associated in particular with the work of the American mathematician Gregory Chaitin and has been presented by him in a large number of popular books and articles.

Chapter 8 contains a discussion of complexity and its relevance to the incompleteness phenomenon in general, together with a presentation of the Paris-Harrington theorem and some comments about the use of axioms of infinity to prove arithmetical statements otherwise unprovable in ZFC.

3

Computability, Formal Systems, and Incompleteness

3.1 Strings of Symbols

The axioms, theorems, and other sentences of a formal system, and also the proofs in the system, can all be taken to be *strings of symbols*. In mathematics and symbolic logic, various special mathematical symbols are used, but this is just a matter of convenience. Here, we may assume that the symbols we are concerned with are those that we find on a standard keyboard and use in ordinary English text—lower and upper case letters, digits, parentheses, and so on—keeping in mind that an empty space, which on a standard keyboard is generated by pressing the space bar, is also a symbol. The definitions and arguments given in the following apply to any starting set of symbols, as long as there are only finitely many symbols in the set.

By a *string* is meant any finite sequence of symbols. For example, the following are strings:

Let me not to the marriage of true minds admit impediments.

aaaaaaaaaaaaaaaaaaaaaaaaaaaaa

=)=)::))

in the nick of time

watashi wa neko da yo

9)#lklK0+==FDBk2++?%

~(Ex)(y)(yEx iff ~yEy)

90239499993930200111O9

As these examples illustrate, strings may or may not be meaningful words, sentences, or other kinds of expressions in some language or other. There is an ambiguity in speaking of "symbols" in connection with strings—should we say that the string "aaaa" contains one symbol (the letter a) or four symbols? We resolve the ambiguity by saying that "aaaa" contains four *occurrences* of a single symbol. The *length* of a string is the number of occurrences of symbols in the string. The examples given are all short strings, but in logic and mathematics we impose no limit on the length of strings, and reason freely about strings that contain more occurrences of symbols than there are grains of sand in the river Ganges. Thus, strings, in this connection, are mathematical objects, just like numbers, and need not have any actual physical representation. Indeed, from a mathematical point of view, strings and natural numbers are interchangeable, in the sense that we can let strings represent numbers or numbers represent strings. A representation of arbitrary strings using numbers, such as will be described in the next section, is what is usually called a *Gödel numbering*. But first let us take a closer look at the more familiar direction, the representation of natural numbers using certain strings.

The strings of digits used in mathematics as well as in ordinary English and many other languages to denote natural numbers (nonnegative integers) will be called *numerical* strings. Thus, the string "165" denotes the number one hundred and sixty-five, and "0" denotes the number zero. If we disallow initial zeros in strings containing more than one digit, every natural number is denoted by a unique corresponding numerical string. In everyday language, we are so used to this way of representing numbers (the decimal positional notation) that it may take an effort to distinguish between the number 165 and the string "165." But even if we consider only the symbols used in English (and not, say, Chinese numerals) there are other common ways of representing the natural numbers as strings. One of them is through the use of counting words, as in "one hundred and sixty-five." In order to extend this notation to cover all of the natural numbers, some new words are needed, and various systems exist. In the system of Landon Curt Noll, the number $3,141,592,653,589,793,238,462,643,383$ is named three octillion, one hundred forty one septillion, five hundred ninety two sextillion, six hundred fifty three quintillion, five hundred eighty nine quadrillion, seven hundred ninety three trillion, two hundred thirty eight billion, four hundred sixty two million, six hundred forty three thousand, three hundred eighty three.

Another way of representing natural numbers as strings that is commonly used in the theory and practice of computing is binary notation (base two positional notation). In binary notation, the above number becomes

1010001001101010101000011111000000001000110111 0
0111110100000010000110100100100011110010110111.

Using strings to denote natural numbers is familiar to us because we have to use strings to denote numbers in order to write down numerical information and when adding and multiplying numbers on paper. The idea of giving arbitrary strings "names" that are natural numbers is less familiar from everyday contexts, but arises naturally in the context of logic and the theory of computability.

It was claimed that not only the theorems and other sentences of a formal language, but also the proofs in a formal system can be taken to be strings of symbols. Since proofs are *sequences* of sentences, this means that we need to be able to represent such sequences too as strings. The easy way of doing this is to choose a symbol that does not occur in any sentence of the language, say £, and use it in strings to separate sentences. Thus for example, the sequence consisting of the sentences "aishiteru no," "suki da yo," "baka da yo" will be represented as the string

aishiteru no£suki da yo£baka da yo.

But what if any symbol can occur in a sentence? It is still possible to represent any sequence of strings as a string, but it is only in Chapter 8 that we will need to consider how to do this.

3.2 Computable Enumerability and Decidability

Computably Enumerable Sets

The set of numerical strings has two properties that are of particular interest in connection with the incompleteness theorem. The first of these properties is that there is a *mechanical procedure* for *generating* the numerical strings, or in other words, for producing and writing them down one after another. We simply first write down the single-digit strings, then systematically write down all two-digit strings (except that we don't include strings beginning with 0), and so on. To fully specify the procedure we need to decide on the order in which strings of the same length are written down, and a natural choice, given our conventional alphabetical ordering of the symbols, is to generate strings of the same length in alphabetical order (in mathematics usually called *lexicographic order*). In this way we obtain

$$0, 1, 2, 3, 4, 5, 6, 7, 8, 9, 10, 11, 12, 13, 14, 15, 16, 17, 18, 19, 20, 21, \ldots$$

As it happens, this particular ordering of the numerical strings is very familiar to us, since it corresponds to the mathematical ordering of the numbers denoted by the strings. The reason for this is twofold. The alphabetical ordering of the digits coincides with the mathematical ordering of the numbers they denote, and we use a positional system in denoting numbers by numerical strings. If we generate the usual names of numbers in words using the same principle of "shorter before longer" combined with lexicographic ordering of strings of the same length, the listing will begin

> one, six, ten, two, five, four, nine, zero, eight, fifty, forty, seven, sixty, three, eighty, eleven, ninety, thirty, twelve, twenty, fifteen, seventy, sixteen, eighteen, fourteen, nineteen, thirteen, fifty-one, fifty-six, fifty-two, forty-one,...

We have described a mechanical procedure for generating numerical strings, such that every numerical string will, if the procedure is continued indefinitely, eventually appear in the list of strings generated. The procedure is *mechanical* in the sense that it calls for no judgment, choice, or problem-solving, but only for the application of a set of explicit rules. To put it differently, it is a simple matter to program a computer to generate numerical strings according to these rules and print them out one after the other. We say that a set E of strings is *computably enumerable* if it is possible to program a computer to compute and print out the members of E, as long as we disregard all limitations of time, energy, and memory. Thus, in this terminology, the set of numerical strings is computably enumerable.

For future arguments, we need to be a bit more specific. When we speak of "programming a computer to compute and print out the members of

E," do we allow repetitions in the output of the program, so that the same member of E may be printed more than once? In fact every infinite set that can be computably enumerated with repetitions can also be computably enumerated without repetitions, and we therefore allow repetitions in the output of the computer, as long as every member of E eventually appears among the strings generated. A more subtle question is what happens if E is finite—does the computer halt when all members of E have been generated, or does the computation continue ad infinitum, even though no new strings will appear in the output? For reasons that are not immediately apparent, in speaking of a computable enumeration of a finite set E, we will in fact allow that a computer programmed to compute and print out the members of E continues this activity forever, and thus either prints out at least one string infinitely many times, or else just whirs silently away doing nothing after producing all of the strings in the set.

The definition of "computably enumerable set" was formulated in terms of sets of strings, but we can also apply it to sets of natural numbers. A set of natural numbers is computably enumerable if and only if the set of corresponding numerical strings is computably enumerable. Similarly, the definition carries over to sets of any mathematical objects that can be represented by strings. For example, every positive rational number has a unique representation as a string m/n where m and n are numerical strings such that the corresponding natural numbers have no common divisor greater than 1, which allows us to speak of computably enumerable sets of positive rational numbers.

It was Alan Turing who, in 1936, introduced a theoretical model of a general-purpose digital computer with unlimited working memory (the universal Turing machine) and made it plausible that everything that can be mechanically computed can be computed by such a machine. Today, Turing computability is one of several equivalent definitions of computability used in logic and mathematics, and we are used to thinking of ordinary computers as physical realizations of the universal Turing machine, except that actual computers have limited memory. In a mathematical treatment of the theory of computability, the Turing machine or some other model of computation is used to give formal definitions of the basic concepts of the theory. In this book we will make do with informal definitions of these basic concepts, which still allow us to understand and appreciate several important results of the theory. All we need to know about the workings of a computer is that executing a program can be described as consisting in a series of *steps*, so that it is possible to carry out a certain number of steps in a computation, do something else for a while, and then return to carry out the next step in the computation.

As an example of the kind of informal reasoning that we can carry out on the basis of these definitions, we see that if A and B are two computably enumerable sets, then their union C, containing the strings belonging to A or B or both, is also computably enumerable. For in order to enumerate C we need only alternately carry out steps in enumerations of A and B, and deliver a member of A or B as output whenever it is generated.

Computably Decidable Sets

A second property of the set of numerical strings is that there is an *algorithm* that allows us to decide, given any string of symbols, whether or not it will *ever* appear in a computable enumeration of the set. An algorithm is a mechanical procedure, such as can be carried out by a computer, which always terminates, generally yielding some string as output. (An alternative terminology is also used in the literature, in which an algorithm is a mechanical procedure that does not necessarily terminate.) The algorithm is simple in this case: we need only check that every symbol in the string is a digit, and that the string does not begin with a zero, unless it has length 1. We say that a set E of strings is *computable* or *computably decidable*, or just *decidable* for short, if it is possible to program a computer to decide, given any string s of symbols, whether or not s is in E. In other words, given s as input, the computer carries out some computation, and then outputs "yes" if s is in E, and "no" if s is not in E. (As in the case of computable enumerability, we can also apply this definition to sets of natural numbers.) A set that is not decidable is said to be (computably) *undecidable*.

Here again, we need to take note of the fact that we are talking about what is mathematically possible, not about what is in any sense feasible. In particular, every *finite* set E of strings is decidable, since an algorithm for deciding whether a string s is in E can consist of simply going through a list of all strings in E, looking for s. A computer can be programmed to carry out this algorithm by including in the program a listing of all the strings in E. Of course if E contains, say, a quadrillion strings, this has nothing to do with what can in fact be done, and it may or may not be possible to actually program (in a different way) a physical computer to decide whether a given string is in E. Statements in the following about what we "can compute" or "can decide" must be understood to carry a similar qualification.

Like the words "complete" and "incomplete," the words "decidable" and "undecidable" are used in logic in two different senses. A sentence A is said to be decidable in a formal system S if either A or its negation is provable in S. Decidability in this sense is a property of sentences and is relative to some formal system. Decidability in the sense introduced here is not a property of sentences, but of sets—sets of strings, of numbers, or of other objects that can be represented as strings or numbers—and it is not relative to any formal system. There is nevertheless a connection between these two senses of "decidable," in that we can use the fact that certain sets are not decidable to prove that certain formal theories have undecidable sentences. To present this result is the main point of this chapter.

Mathematically Definable Sets

To ask whether a set of strings is computably enumerable, or computably decidable, makes sense only if the set can be given a formal (mathematical) definition. For example, is the set of sentences in English decidable? Here, we need to specify just what is meant by a sentence in English. If we restrict ourselves to sentences that have been spoken or written, or will be spoken or written in the future, it is clear that there is a limit on their length, since it will never happen that a sentence containing billions of symbols is spoken or written. So, since every finite set of strings is decidable, does it follow that the set of English sentences of length smaller than a billion is decidable? No, for it is simply not determinate which strings belong to this supposed set, which means that no set has yet been specified. A natural language like English, as opposed to the formal languages studied in logic, is not defined through mathematical rules, but through actual usage, and as soon as we consider what users of English accept as an English sentence, we find variations of time and place, variations among different speakers, cases of people changing their mind, and cases where nobody can tell whether a string is a sentence or not.

This does not mean that the question whether the set of sentences in English is decidable is meaningless. But to make sense of it, we need to introduce a theoretical model of the English language in the form of a formal grammar that can reasonably be held to make explicit rules for constructing English sentences that agree with how the language is in fact used. This formal grammar will allow an infinite number of strings as

English sentences, and it becomes a nontrivial question whether the set of sentences is decidable.

In logic and mathematics, the sets of strings that we investigate are defined at the outset in mathematical terms, for example, as the set of sentences derivable from certain axioms using certain formal rules of inference. The question whether the set is computably enumerable or decidable thus always makes good sense.

A Gödel Numbering

Let us first observe that the procedure described for generating the numerical strings can be modified so as to generate *all* strings. We first generate the strings consisting of only one symbol (there are only finitely many of these), and then systematically generate, in lexicographic order, the strings consisting of two symbols (again, there are only finitely many), and so on.

Thus suppose our symbols are the symbols in positions 32–126 in the ASCII table. These symbols, in alphabetical order, are space, !, “, #, $, %, &, ‘, (,), *, +, comma, -, period, /, the digits 0–9, :, ;, <, =, >, ?, @, the letters A–Z, [, \,], ^, _, ‘, the letters a–z, |, }, ∼. In generating all strings we start by generating the 95 one-symbol strings, then the $95 \times 95 = 9025$ two-symbol strings in lexicographic order (the first being the string consisting of two occurrences of space), and so on.

This also gives us an example of a *Gödel numbering*, a way of representing arbitrary strings, not just numerical ones, by numbers. We simply associate every string with its position in the enumeration of all strings. Thus the 95 one-symbol strings are given Gödel numbers 1–95, and the following 9025 two-symbol strings will have Gödel numbers 96–9120. The number 0 is assigned to the *empty string*, like the empty set a convenient mathematical construct, which contains no symbols at all.

This Gödel numbering is only one of infinitely many ways of representing arbitrary strings by numbers. It satisfies two essential conditions. First, given a string we can mechanically compute its Gödel number, by simply generating strings and counting them until we come to the string whose Gödel number we seek. Second, given a number, we can decide by a similar computation whether it is the Gödel number of a string—which in this particular numbering is true of every number—and if so which one. As a consequence, a set of strings is computably enumerable or decidable if and only if the set of Gödel numbers of strings in the set is computably enumerable or decidable.

3.3 Undecidable Sets

Two Basic Connections

There are two basic relations between computable enumerability and computable decidability that are the key to many observations involving these concepts. The first basic connection is the following:

> Every computably decidable set is computably enumerable.

This is because we can always (in theory) program a computer to generate the strings in a decidable set E by generating *every* string internally, but retaining and printing out a string only if a computation shows it to be in E, and discarding it if it is not in E.

In practice, if a set is decidable, there are usually more simple and direct ways of generating its members. For example, to generate the numerical strings, we don't systematically generate all strings and sift out those that are numerical, but instead systematically generate all strings from the restricted alphabet containing only the symbols 0–9, skipping strings with initial zeros, except for the one-symbol string consisting only of 0. But the general procedure described is always available, and it shows that every decidable set can be computably enumerated.

The *complement* of a set E of strings is the set of all strings that are not in E. The second basic observation is that

> A set E is computably decidable if and only if both E and its complement are computably enumerable.

To see this we first note that if E is computably decidable, so is the complement of E, so it follows from the first basic fact that both E and its complement are computably enumerable. If E and its complement are both computably enumerable, a computer can decide whether a given string is in E as follows. First carry out one step in a computable enumeration of E and one step in a computable enumeration of the complement of E. If the string has not yet appeared in either set, carry out another step in both computations. Eventually the string will appear in either E or the complement of E, and the computer outputs "yes" or "no" accordingly.

The Undecidability Theorem

We can now state the starting point of the theory of computably enumerable sets:

Undecidability theorem (Turing, Church). *There are computably enumerable sets which are not computably decidable.*

Before considering proofs that such sets exist and examples of such sets, it is worthwhile to ponder their general characteristics. Suppose E is a computably enumerable set that is not decidable. There is then an algorithm for systematically generating all of the strings in the set. Thus if s is a member of E, it is possible, in theory at least, to show that s is a member of E, by generating members of E until s appears. (Computably enumerable sets have sometimes been called *semi-decidable.*) However, there is no mechanical procedure which, applied to an arbitrary string s, will always give information about whether or not s is in E. Thus, if s is *not* a member of E, this can only be shown, if at all, by special reasoning. In proving the first incompleteness theorem using an undecidable computably enumerable set E, we will see that unless a formal system S incorrectly decides some statements of the form "k is in E," there will be statements of the form "k is in E" that are undecidable in S.

Turing's Proof of the Undecidability Theorem

Turing proved the undecidability theorem by proving the "recursive unsolvability of the halting problem." Here we need to take note of some traditional terminology often used in the literature. "Recursively enumerable" or "effectively enumerable" are both synonyms for "computably enumerable." The term "recursively unsolvable problem" does not refer to a problem in the sense of an open question, but to a *set* of problems. A recursively unsolvable problem is given by a set E (of natural numbers, of strings, or of some other kind of mathematical objects representable as numbers or strings) that is not decidable. The set of problems of the form "is a in E?" is computably unsolvable in the sense that there is no mechanical procedure that applied to *any* problem of this form will yield a correct answer (yes or no).

To understand the connection between the undecidability theorem and the first incompleteness theorem, it is not necessary to study any proof of the undecidability theorem, but a presentation of Turing's proof will be given here for the interested reader.

Let us fix a programming language—Basic, C, Java, or any other standard language, since they are all equivalent from the point of view of computability theory. A reader who is not familiar with any programming language need only think of it as a language for expressing instructions on how to proceed in a mechanical step-by-step computation. The computation may require some initial input strings to work with, and it may or may not eventually terminate, returning some output string. A program in the language is itself a string of symbols. The rules for writings programs are such that the set of programs is decidable, and so is the set of programs that expect one string as input, and when executed either deliver a string as output or else never terminate. Thus, there is a computable enumeration of these (infinitely many) programs:

$$P_0, P_1, P_2, \ldots$$

We say that the number i is the *index* of the program P_i.

Now let K be the set of i such that P_i terminates and outputs a value when given the (numerical string denoting the) index i of the program itself as input. This somewhat oddly defined set K is in fact computably enumerable. For the verification of this, all we need to know about the programs P_i is that their computations proceed in a step-by-step fashion so that the set of true statements of the form

> The computation carried out by P_i with input k terminates after at most n steps,

where k and n are numerical strings, is decidable. This is so because given i, k, and n, we can first generate the program P_i in the enumeration of programs, and then start P_i on its computation with input k, but allow it to proceed for at most n steps. Now to computably enumerate the elements of K, we just go through all strings of the indicated form, and whenever we encounter a true such statement with $k = i$, we add i to the listing of the elements of K.

So K is computably enumerable, but it is not decidable. For suppose K is decidable. We can then define a procedure which given any input i first checks whether i is in K. If not, we give 0 as output. If i is in K, so that P_i does terminate with i as input, we let P_i compute its result and then give as output that result with a further symbol added at the end. Since this defines a program P that given any string computes another string as output, P must be identical with P_m for some m. But P and P_m do not agree on m, so they are not identical. Hence, K is not decidable.

Turing's argument is what is known in logic as a *diagonalization argument.* In its most basic form, the diagonalization argument goes as follows.

Given a relation $R(a, b)$ between elements a and b in some set A, if we define the property P of elements in A to hold if and only if $R(a, a)$ does not hold, this property P is not identical with the property R_a for any a, where $R_a(b)$ is defined to hold if and only if $R(a, b)$ holds. Variations of this very simple argument have been used in set theory and logic since it was first introduced in the late nineteenth century by the German mathematician Georg Cantor, the creator of set theory. Some form of diagonalization argument lies at the basis of most proofs, or perhaps of every proof, of the undecidability theorem and of the first incompleteness theorem, when the incompleteness theorem is given a proof that implies the existence of undecidable *Goldbach-like* statements. (See Section 2.1 for a definition of the Goldbach-like statements.)

Two further examples of effectively enumerable but undecidable sets will be introduced, *simple* sets as defined by Post and the set of solvable Diophantine equations.

Post's Simple Sets

Emil Post, who did pioneering work in the theory of computably enumerable sets, came up with a category of computably enumerable sets which are in a sense extremely undecidable. A *simple* set is a computably enumerable set A of natural numbers such that the complement of A, although infinite, has no infinite computably enumerable subset. Thus, although there are infinitely many numbers not in A, it is not possible for any mechanical procedure to generate more than finitely many of them. The set K defined in Turing's proof is not simple, for we can easily construct infinitely many programs that do not terminate for any input.

That there are simple sets is not obvious. Post showed how to define a simple set using an enumeration of the computably enumerable sets. In Chapter 8, a specific example of a simple set, which was found some 20 years later, will be defined, the set of *compressible* strings.

Hilbert's Tenth Problem and the MRDP Theorem

Number ten on the list of 23 problems put forward by David Hilbert in the year 1900 was the problem of finding an *algorithm* for deciding whether or not a Diophantine equation (see Section 2.1) has *any* solution. In 1970 it was established, as a consequence of what is known as the *Matiyasevich-Robinson-Davis-Putnam* (MRDP) theorem, that there is no such algorithm.

To simplify matters, we will restrict the solutions of Diophantine equations to *nonnegative* integers. (It can be shown that the question whether a Diophantine equation has any solution in integers can always be reduced to the question whether another Diophantine equation has a solution in nonnegative integers.) Thus, a Diophantine equation $D(x_1, \ldots, x_n) = 0$ is said to be *solvable* if it has a solution $k_1, \ldots, k_n$ in natural numbers.

We first note that the set of solvable Diophantine equations is computably enumerable. To see this, note that the set of all statements (true or false) of the form

$$D(x_1, \ldots, x_n) = 0 \text{ has the solution } x_1 = k_1, \ldots, x_n = k_n$$

where $k_1, \ldots, k_n$ are numerical strings is decidable, and hence computably enumerable. So, in order to generate all solvable Diophantine equations, we need only generate all strings of this form, and for each string check whether $D(x_1, \ldots, x_n)$ does in fact evaluate to 0 when $x_1 = k_1, \ldots, x_n = k_n$. If this is the case, we list $D(x_1, \ldots, x_n) = 0$ as a solvable Diophantine equation.

Similarly, for any given Diophantine equation $D(x_1, \ldots, x_n, y) = 0$, the set E_D of k for which there exists a solution of the equation with $y = k$ is computably enumerable. This is so since we can generate the set of solvable equations of the form $D(x_1, \ldots, x_n, k) = 0$, obtained by replacing the unknown y with the number k, and list k as a member of E_D for every such equation found. A computably enumerable set that is equal to E_D for some D is said to be *Diophantine*.

The MRDP theorem states that *every* computably enumerable set of natural numbers is Diophantine. One consequence is that since there are computably enumerable sets that are not decidable, there cannot be any algorithm for deciding whether or not a given Diophantine equation has any solution. Thus, the set of solvable Diophantine equations is an example of a computably enumerable but not decidable set, and the set of *unsolvable* Diophantine equations is not computably enumerable.

The MRDP theorem has another consequence, which depends on the fact that the theorem can be proved in elementary arithmetic, by a proof formalizable in PA. Given any Goldbach-like statement A, as informally defined in Section 2.1 (and formally in the Appendix), we can construct a corresponding Diophantine equation $D(x_1, \ldots, x_n) = 0$ such that it is provable in PA that A is true if and only if the equation has no solution. In particular, such an equation can be found for the statement "PA is consistent."

3.4 Computability and the First Incompleteness Theorem

Formal Systems

A formal system is determined by a formal language, a set of inference rules, and a set of axioms. The set of sentences of the language is assumed to be a decidable set of strings. The axioms (which are certain sentences in the language of the system) and the inference rules need to be defined in such a way that the system satisfies the following:

Basic property of formal systems. *The set of theorems of a formal system is computably enumerable.*

For example, in the case of a first-order theory, it is possible to choose the set of axioms (including "purely logical" axioms, which are axioms in every first-order theory) to be decidable, and to have only one rule of inference, whereby B follows from A together with "if A then B." A proof in such a system is a finite sequence of sentences $A_1, \ldots, A_n$ where for each i, either A_i is an axiom or there are earlier sentences A_j and "if A_j then A_i" in the sequence. Thus, the set of proofs is also decidable, and as a consequence the basic property holds, since to computably enumerate the theorems of the system we need only go through all strings to check if they are proofs and for each string that is a proof output the last sentence in the sequence. In general, it is sufficient that the set of proofs is computably enumerable for the basic property to hold, since we need only systematically generate proofs and pick out their conclusions in order to generate the theorems. Conversely, if the basic property holds, we can always define proofs in the system so as to make the set of proofs computably enumerable.

The basic property implies that it is always possible to *search* for a proof of a given sentence A in a mechanical way, and if there is any such proof, it will eventually be found. If A is not a theorem of the system, the search will just go on forever. In practice, when proving theorems, we do not usually search mechanically, but are guided by more or less clear ideas about how to prove the theorem. In the field of *automatic theorem proving*, one tries to program computers to search for proofs "intelligently," rather than just by systematic brute force, although such automated "intelligent" searching for proofs is still mechanical in the sense that it follows a specifiable set of rules. When we look for proofs informally, we may or may not be following rules that computers can be programmed to follow, but we are not usually able to specify any such rules.

A formal system is *decidable* if the set of theorems of the system is not only computably enumerable, but also decidable. A complete formal system is always decidable. For either it is inconsistent, in which case the set of theorems is identical with the set of sentences, or else we can decide whether a sentence A is a theorem by enumerating the theorems of the system until we come upon either A or not-A. Thus, for example, the elementary theory of the real numbers (see Section 2.3), which is complete, is also decidable. (This is not to say that there is any algorithm which can be used in practice to decide whether an arbitrary sentence in the language of the theory is provable.) There are also theories that are decidable without being complete.

The First Incompleteness Theorem

Suppose S is a formal system that contains enough arithmetic to be able to prove all true statements of the form

$$D(x_1, \ldots, x_n) = 0 \text{ has the solution } x_1 = k_1, \ldots, x_n = k_n.$$

Not a great deal of arithmetic is needed to prove every true statement of this form—we only need to be able to carry out multiplication, addition, and subtraction of integers, and to handle the logic of $=$.

Now consider the theorems of S of the form

$$D(x_1, \ldots, x_n) = 0 \text{ has no solution.}$$

If S is consistent, every such theorem of S is true. For if the equation does have a solution, this is provable in S, by the assumption on S, so it cannot also be provable in S that the equation has no solution. So the set of all equations for which it is provable in S that they have no solution is a computably enumerable subset of the set of equations that do not have a solution. But since the latter set is *not* computably enumerable, it follows that there are infinitely many equations $D(x_1, \ldots, x_n) = 0$ that have no solution, but for which this fact is not provable in S. And if we further assume that S is Σ-sound, and thus does not prove any *false* statements of the form "The Diophantine equation $D(x_1, \ldots, x_n) = 0$ has at least one solution," it follows that there are infinitely many equations $D(x_1, \ldots, x_n) = 0$ for which it is undecidable in S whether or not they have any solution.

We can make a stronger statement. The set A of all equations for which it is provable in S that they have no solution is a computably enumerable subset of the set B of equations that do not have a solution. If the set $B\setminus A$ of elements in B that do not belong to A were computably enumerable, B would be the union of two computably enumerable sets and therefore itself computably enumerable. Thus, the set of true statements of the form "The Diophantine equation $D(x_1,\ldots,x_n) = 0$ has no solution" that are not provable in S is not only infinite, but is not computably enumerable.

This version of the first incompleteness theorem based on computability theory shows that the use of (arithmetical formalizations of) self-referential sentences is not essential when proving incompleteness. Indeed, there is no assumption in the proof that S is capable of formalizing self-reference. It also shows that the content and structure of S, beyond encompassing some elementary arithmetic, is not relevant if we only want to establish incompleteness. In particular, whether S is a first-order theory or uses what is called second-order logic makes no difference, as long as S satisfies the basic property of formal systems.

The same argument can be carried through using any computably enumerable but undecidable set E instead of the set of solvable Diophantine equations. What is not obvious is that all statements of the form "k is in E" can be formulated in the language of arithmetic for any computably enumerable E, and all true statements of this form can be proved on the basis of some basic principles of arithmetic. To make it clear that this is in fact the case, we need a formal specification of some model of computation (such as Turing machines), and we also need to use an arithmetization of the language of computation, similar to the arithmetization of syntax used in Gödel's proof of the first incompleteness theorem. Using the set of solvable Diophantine equations as E has the advantage that no arithmetization of syntax is needed, and only simple arithmetical statements are involved.

The version of the first incompleteness theorem obtained through this argument has the same strength as Gödel's original version: we need to assume that S is Σ-sound (does not prove any false Goldbach-like statement), which is equivalent to assuming that S does not prove any false statement of the form "The Diophantine equation $D(x_1,\ldots,x_n) = 0$ has at least one solution." For some computably enumerable but undecidable sets E, it is perfectly possible for a consistent system to decide all statements of the form "n is in E," although infinitely many statements will be decided incorrectly. In the special case of the set of unsolvable Diophantine equations, however, no consistent system (incorporating a certain amount of arithmetic) settles every statement of the form "The Diophantine equation

$D(x_1, \ldots, x_n) = 0$ has at least one solution," since the undecidable Rosser sentence (see Section 2.7) is equivalent to such a statement.

Suppose we choose E as a *simple* set in Post's sense. Then S can prove only *finitely* many statements of the form "k is not in E," although there are infinitely many true such statements. We will consider this further in connection with Chaitin's incompleteness theorem in Chapter 8.

For a final illustration of the power of the basic ideas of the theory of computability, we turn to another topic first commented on by Gödel.

Essential Undecidability and Speeding Up Proofs

Let us look at the above argument from a different point of view. Given that S is Σ-sound, it follows that a Diophantine equation $D(x_1, \ldots, x_n) = 0$ has at least one solution if and only if S proves the statement "The Diophantine equation $D(x_1, \ldots, x_n) = 0$ has at least one solution." The set of theorems of S therefore cannot be decidable, since otherwise there would be an algorithm for deciding whether or not a Diophantine equation has a solution—just check whether the corresponding statement is provable in S.

It follows in particular that PA, and every Σ-sound theory that contains PA as a part, which we call an *extension* of PA, is undecidable. By a variant of this argument, it can be shown that in fact every *consistent* extension of PA is undecidable. We say that PA is *essentially* undecidable.

To prove a theory T undecidable is another way of proving that it is incomplete, since, as has been noted, a complete formal system is decidable. Suppose T is essentially undecidable, and let A be a sentence undecidable in T. We can now observe (assuming T to incorporate the logic of "not" and "or") that the set M of theorems of $T' = T + A$ that are not theorems of T is not computably enumerable. For given any sentence B, the sentence "A or B" (which is a theorem of T') is a member of M if and only if "A or B" is not a theorem of T, which is to say if and only if B is not a theorem of $T +$ not-A. So if M is effectively enumerable, the set of non-theorems of $T +$ not-A is effectively enumerable, implying that $T +$ not-A is decidable, which is inconsistent with our two assumptions that A is undecidable in T and that every consistent extension of T is undecidable.

The argument in the preceding paragraph concerned the richness of the set of new theorems provable in $T+A$ but not in T. Essential undecidability has a further consequence, concerning theorems of $T + A$ that are *also* theorems of T, which although easily established is by no means obvious.

Suppose we extend PA by a statement A undecidable in PA to the theory PA′. If we consider theorems of PA and how they can be proved in PA′, is it necessarily the case that there is some theorem of PA which can be proved more efficiently in PA′, in the sense that it has a *shorter* proof in PA′ than any proof that can be found in PA? Here, by a shorter proof we mean one with a smaller Gödel number. The answer is yes, and in fact there must be theorems of PA that have "much" shorter proofs in PA′. For example, there is a theorem A of PA for which $10^{1000} \times p' < p$, a theorem A for which $2^{p'} < p$, and so on, where p' is the length of the shortest proof of A in PA′ and p is the length of the shortest proof of A in PA. In fact, for *any* computable function f that takes natural numbers to natural numbers, there is a sentence A for which $f(p') < p$. (Gödel was the first to state a result of this kind, in 1936.)

To see this, suppose that on the contrary there is a computable f such that there is no theorem A of PA for which $f(p') < p$. Then for every A, if A is a theorem of PA′ with a proof of length n, A is a theorem of PA if and only if A has a proof in PA of length smaller than $f(n)$. This means that we have a procedure for effectively enumerating those theorems of PA′ that are *not* theorems of PA, which contradicts the essential undecidability of PA.

4

Incompleteness Everywhere

4.1 The Incompleteness Theorem Outside Mathematics

The incompleteness theorem is a theorem about the consistency and completeness of formal systems. As noted in the introductory chapter, "consistent," "inconsistent," "complete," "incomplete," and "system" are words used not only in a technical sense in logic, but in various senses in ordinary language, and so it is not surprising that the incompleteness theorem has been thought to have a great many applications outside mathematics.

A few examples:

- Religious people claim that all answers are found in the Bible or in whatever text they use. That means the Bible is a complete system, so Gödel seems to indicate it cannot be true. And the same may be said of any religion which claims, as they all do, a final set of answers.

- As Gödel demonstrated, all consistent formal systems are incomplete, and all complete formal systems are inconsistent. The U.S. Constitution is a formal system, after a fashion. The Founders made the choice of incompleteness over inconsistency, and the Judicial Branch exists to close that gap of incompleteness.

- Gödel demonstrated that any axiomatic system must be either incomplete or inconsistent, and inasmuch as Ayn Rand's philosophy of Objectivism claims to be a system of axioms and propositions, one of those two conditions must apply.

It will be noted that all of these misstate the incompleteness theorem by leaving out the essential condition that the system must be capable of formalizing a certain amount of arithmetic. There are many complete and consistent formal systems that do not satisfy this condition. If we remember to include the condition, supposed applications of the incompleteness theorem such as those illustrated will less readily suggest themselves, since neither the Bible, nor the constitution of the United States, nor again the philosophy of Ayn Rand is naturally thought of as a source of arithmetical theorems.

The incompleteness theorem is a mathematical theorem, dealing with formal systems such as the axiomatic theory PA of arithmetic and axiomatic set theory ZFC. Formal systems are characterized by a formal language, a set of axioms in that language, and a set of formal inference rules which together with the axioms determine the set of theorems of the system. The Bible is not a formal system. To spell this out: it has no formal language, but is a collection of texts in ordinary language, whether Latin, English, Japanese, Swahili, Greek, or some other language. It has no axioms, no rules of inference, and no theorems. Whether something follows from what is said in the Bible is not a mathematical question, but a question of judgment, interpretation, belief, opinion. Similarly for the Constitution and the philosophy of Ayn Rand. Deciding what does or does not follow from these texts is not a task for mathematicians or computers, but for theologians, believers, the Supreme Court, and just plain readers, who must often decide for themselves how to interpret the text and have no formal rules of inference to fall back on.

Thus, we need only ask the question "is the Bible (the Constitution, etc.) a formal system?" for the answer to be obvious. Of course it's not a formal system. It's nothing like a formal system. To be sure, if we set aside the mathematical notion of a formal system and use the words "formal" and "system" in an everyday sense, it can be said that the Constitution is a formal system "after a fashion." That is to say, from the point of view of ordinary language, the Constitution contains lots of formal language, and it looks rather systematic. But nobody would seriously claim that there is any such thing as the formally defined language, the axioms, and the rules of inference of the Constitution. So the incompleteness theorem does not apply to the Constitution, the Bible, the philosophy of Ayn Rand, and so on.

Without pretending that these various systems of thought, legislation, philosophy, and so on constitute formal systems, we could apply to them *analogues* of the formal notions of consistency and completeness. For the Bible to be complete in such a sense, analogous to that used in logic, would mean that every statement that makes sense in the context of Bible reading could be reasonably held to be decided by the Bible, in the sense that either that statement or its negation can be held to be explicitly or implicitly asserted in the Bible. If we raise the question whether the Bible is complete in this sense, the answer is again pretty obvious: it is not. For example, it makes good sense in the context of Bible reading to ask whether Moses sneezed on his fifth birthday. No information on this point can be found in the Bible. Hence, the Bible is incomplete. Similarly, the Constitution is incomplete, since it does not tell us whether or not wearing a polka-dot suit is allowed in Congress. Ayn Rand's philosophy of Objectivism is incomplete, since we cannot derive from it whether or not there is life on Mars, even though it makes sense in the context of Objectivism to ask whether there is life on Mars. We don't need Gödel to tell us that these "systems" are in this sense incomplete. Trivially, any doctrine, theory, or canon is incomplete in this analogical sense. Such trivial observations are presumably not at issue in the comments quoted. But there is no more substantial or interesting use to be made of the incompleteness theorem in discussing the Bible, the Constitution, Objectivism, etc.

Similar remarks apply to the following reflections by John Edwards in the electronic magazine *Ceteris Paribus*:

> We can view rules for living, whether they are cultural mores of the sort encoded in maxims, or laws, principles, and policies meant to dictate acceptable actions and procedures, as axioms in a logical system. Candidate actions can be thought of as propositions. A proposition is proved if the action it corresponds to can be shown to be allowed or legal or admissible within the system of rules; it is disproved if it can be shown to be forbidden, illegal, or inadmissible. In the light of Gödel's theorem, does it not seem likely that any system of laws must be either inconsistent or incomplete?

To say that we "can view" rules for living and so on as axioms in a logical system is unexceptionable since anything, broadly speaking, can be viewed as anything else. Furthermore, in the case of viewing a lot of things as "systems" with "axioms" and "theorems," it is demonstrably the case that many people find it natural and satisfying to view things this way. It is a different question whether any conclusions can be drawn, or any substantial claims supported, on the basis of such analogies and metaphors. In the quoted passage, the suggested conclusion is that a system of laws must be "inconsistent or incomplete." Given the accompanying explanation of what "inconsistent" and "incomplete" mean here, it is an easy observation that all systems of laws, rules of living, and so on, are both inconsistent and incomplete and will remain so. In other words, in the case of legal systems, there will always be actions and procedures about which the law has nothing to say, and there will always be actions and procedures on which conflicting legal viewpoints can be brought to bear. Hence the need for courts and legal decisions. References to Gödel's theorem can only add a rhetorical flourish to this simple observation.

4.2 "Human Thought" and the Incompleteness Theorem

A prominent example of a "system" to which the incompleteness theorem is sometimes thought to be applicable is "human thought," also known as "the human mind," "the human brain," or "the human intellect." Here the term "human thought" will be used. The same general comments apply as in the case of supposed applications of the theorem to the Bible, the Constitution, the philosophy of Ayn Rand—there is no such thing as the formally defined language, the axioms, and the rules of inference of "human thought," and so it makes no sense to speak of applying the incompleteness theorem to "human thought." There are however special factors that influence this particular invocation of the incompleteness theorem and may seem to lend the view that the incompleteness theorem applies to "human thought" a certain appeal. One formulation of such a view is the following:

> Insofar as humans attempt to be logical, their thoughts form a formal system and are necessarily bound by Gödel's theorem.

Here by "attempt to be logical" one could perhaps mean "try to argue in such a way that all conclusions reached are formal logical consequences of a specified set of basic assumptions stated in a formal language." Within certain narrowly delimited areas of thought, this description is indeed applicable. For example, in attempting to factor large numbers, the conclusions reached, of the form "the natural number n has the factor m," are intended to be logical consequences of some basic arithmetical principles. This does not mean, however, that one is restricted to applying logical rules starting from those basic principles in arriving at the conclusion. On the contrary, any mathematical methods may be used, and in the case of factorization, there would be nothing wrong with using divination in tea leaves if this turned out to be useful, since it can be checked whether a claim of the form "the natural number n has the factor m" is true or not (and thus whether it is a logical consequence of basic principles). On a stricter interpretation, "attempting to be logical" might consist in the actual application by hand of certain specific formal rules of reasoning, as when we carry out long division on paper, and then indeed it may be reasonable to say that our thinking is closely associated with a formal system. But these are not typical intellectual activities. If by "attempting to be logical" one means only, as in the everyday meaning of "logical," attempting to make sense, to be consistent, not jump to conclusions, and so on, we can't point to any formal systems that have any particular relevance to our thinking. Formal systems are studied and applied in mathematical contexts and in programming computers, not in political debates, in legal arguments, in formal or informal discussions about sports, science, the news, or the weather, in problem solving in everyday life, or in the laboratory. And even when we strive to be as mindlessly computational as we can, to say that "our thoughts form a formal system" is metaphorical at best.

A line of thought that is often put forward in this connection is the following. Let us accept that the incompleteness theorem can only be sensibly applied when we are talking about proving mathematical statements, and not in the context of Bible reading, philosophical or political discussion, and so on. But human beings *do* prove mathematical, and in particular arithmetical, statements. If we say that "human thought," when it comes to proving arithmetical statements, is *not* bound by the incompleteness theorem, are we then not obliged to hold that there is something essentially noncomputable about human mathematical thinking which allows it to transcend the limitations of computers and formal systems, and perhaps even some irreducibly spiritual, nonmaterial component of the human mind? Such a conclusion is as welcome to many as it is unwelcome to others, and this tends to influence how people view attempts to apply the incompleteness theorem to human thought.

The view here to be argued is the following. It doesn't matter, when we talk about the incompleteness theorem and its applicability to human thought, whether people are similar to Lieutenant Commander Data of *Star Trek* fame, with "positronic brains" whirring away in their heads, influenced from time to time by their "emotion chips," or are on the contrary irreducibly spiritual creatures transcending all mechanisms.

The basic assumption made in disputes over whether or not human thought is constrained by the incompleteness theorem is that we can sensibly speak about "what the human mind can prove" in arithmetic. Assuming this, we may ask whether the set M of all "humanly provable" arithmetical sentences is computably enumerable. If it is, "the human mind" is subject to Gödel's incompleteness theorem, and there are arithmetical statements that are not "humanly decidable" (given that "human thought" is consistent). If M is not computably enumerable, human thought surpasses the powers of any formal system and is in this sense not constrained by the incompleteness theorem. To ask whether or not M is computably enumerable is, given this assumption, to pose a challenging and highly significant problem, a solution to which would be bound to be enormously interesting and illuminating.

But is there such a set M of the "humanly provable" arithmetical statements? This assumption is implausible on the face of it and has very little to support it. Actual human minds are of course constrained by many factors, such as the limited amount of time and energy at their disposal, so in considering what is "theoretically possible" or "possible in principle" for "the human mind" to prove, some theory or principle of the workings of "the human mind" is presupposed, one that allows us to speak in a theoretical way of the "humanly provable" arithmetical statements. What theory or set of principles this might be is unclear, and the fact of the variability and malleability of "the human mind" makes it highly unlikely that any such theory is to be had.

Let us consider the question what "the human mind" *has* proved. Some human minds reject infinitistic set theory as meaningless, whereas others find it highly convincing and intuitive, with a corresponding sharp disagreement over whether or not certain consistency statements have been proved or made plausible. The question what "the human mind can prove" presupposes an agreement on what is or is not a proof. Lacking any theoretical characterization of all possible proofs, and even any general agreement on what existing arguments *are* proofs, we can only ask what "the human mind can prove" using certain formally specified methods of reasoning, in which case the human mind becomes irrelevant, and the question is one about what is provable in certain formal systems.

As for malleability, Errett Bishop, who supported and worked in what is known as constructive mathematics, spoke of "the inevitable day when constructive mathematics will be the accepted norm." From what we know about the human mind, it is perfectly conceivable that such an event could take place. Indeed, we know nothing to rule out the possibility that the accepted mathematical norm will in the future be such that even Bishop's constructive mathematics is regarded as partly unjustifiable. It is equally conceivable that people can convince themselves, in one way or another, of the acceptability of extremely nonconstructive principles that are today not considered evident by anybody. In this sense, then, there may be no limit at all on the "capacity of the human mind" for proving theorems, but of course there is nothing to exclude the possibility that false statements will be regarded as proved because principles that are not in fact arithmetically sound will come to be regarded as evident.

The question of the actual or potential reach of the human mind when it comes to proving theorems in arithmetic is not like the question how high it is possible for humans to jump, or how many hot dogs a human can eat in five minutes, or how many decimals of π it is possible for a human to memorize, or how far into space humanity can travel. It is more like the question how many hot dogs a human can eat in five minutes without making a totally disgusting spectacle of himself, a question that will be answered differently at different times, in different societies, by different people. We simply don't have the necessary tools to be able to sensibly pose large theoretical questions about what can be proved by "the human mind." This point will be considered again in Chapter 6, in connection with Gödelian arguments in the philosophy of mind.

There is yet another approach to the application of the incompleteness theorem to human thought, which does not seek undecidable statements either in ordinary informal reasoning or in mathematics, but suggests that Gödel's *proof* of the first incompleteness theorem can be carried through in nonmathematical contexts. This suggestion will be considered next.

4.3 Generalized Gödel Sentences

Mathematical

As mentioned in Chapter 2, the technique of introducing provable fixpoints that Gödel invented has been used since in very many arguments in logic, not just in the proof of the first incompleteness theorem. An example follows.

The basic property of a formal system, as emphasized in Section 3.4, is that its set of theorems is computably enumerable. If we no longer require this property, but retain the formal language, the set of axioms, and the rules of inference, we get the concept of a formal theory in a wider sense, also studied in logic. A theory in this generalized sense need not be relevant to mathematical knowledge, since we may not have any method for deciding whether something is a proof in the theory or not, but the concept is very useful in logical studies.

Using this more liberal notion of "theory," we can define T as the theory obtained by adding *every* true statement of the form "the Diophantine equation $D(x_1, \ldots, x_n) = 0$ has no solution" to PA as an axiom. The set of Gödel numbers of true statements of this form can be defined in the language of arithmetic, so using the fixpoint construction we can formulate an arithmetical statement A for which it is provable in PA that A is true if and only if it is not provable in T. (This A will not be a Goldbach-like statement.) If A is false, then it is provable in T, which is impossible since all axioms of T are true. So A is in fact true but unprovable in T.

In this way, the fixpoint construction used in Gödel's proof can be extended to show that many theories that do not constitute formal systems are still incomplete. But suppose we add *every* true arithmetical sentence as an axiom to PA. The resulting theory T (known as "true arithmetic") is obviously complete, so where does the argument break down in this case? It breaks down because we can no longer form a "Gödel sentence" for T, since (as this very argument shows) the property of being the Gödel number of a true arithmetical sentence cannot be defined in the language of arithmetic.

Nonmathematical

So why not, instead of seeking to apply the incompleteness theorem to nonmathematical systems, just mimic Gödel's proof of the theorem by formulating a "Gödel sentence" for such systems? Thus, we might come up with the following:

- The truth of this sentence cannot be established on the basis of the Bible.

- The truth of this sentence cannot be inferred from anything in the Constitution.

- This sentence cannot be shown to be true on the basis of Ayn Rand's philosophy.

Before considering these dubious statements, we need to dispose of a particular objection that is often raised against them, regarding the use of the phrase "this sentence" to denote a sentence in which the phrase occurs. Such self-reference has been thought to be suspect for various reasons. In particular, the charge of leading to an infinite regress when one attempts to understand the statement has been leveled against it. Such a charge makes good sense if one takes "this" to refer to what is traditionally called a "proposition" in philosophy, the *content* of a meaningful sentence. But in the above statements, as in the corresponding construction in the original Gödel sentence, "this sentence" refers to a syntactic object, a sequence of symbols, and there is no infinite regress involved in establishing the reference of the phrase. As in the case of the Gödel sentence, we can make this fact apparent by eliminating the phrase "this sentence" in favor of the use of syntactical operations, such as substitution or "quining" (see Section 2.7), but we can also simply replace the first sentence with

> The truth of the sentence P cannot be established on the basis of the Bible

together with the explicit definition

> $P =$ "The truth of the sentence P cannot be established on the basis of the Bible."

This explicit definition is as unproblematic as any in logic or mathematics, since we are simply introducing a symbol to denote a particular string of symbols. (In particular, the definition of P does not presuppose that the string of symbols on the right means anything.) In the following, the shorter version of self-referential sentences, containing "this sentence," will be used.

Now, in considering the three proposed "Gödel sentences," one's first impulse may be to regard them as unproblematically true, but not as presenting any inadequacy of either the Bible, the Constitution, or Ayn Rand's philosophy. After all, it is by no means obvious that the statements even make sense in the context of Bible studies, etc. The Gödel sentence for PA is in a different case, since that sentence is itself an arithmetical sentence, and the inability of PA to settle it indicates a gap in the mathematical power of PA. That the Constitution (let us suppose) does not suffice to establish the truth of "The truth of this sentence cannot be inferred from anything in the Constitution" indicates no gap whatsoever in the Constitution, which after all was not designed as an instrument for the discussion of logical puzzles.

However, the three statements are not as unproblematic as they may seem, since we might as well go on to formulate a more far-reaching "Gödel sentence" of this kind:

> This sentence cannot be shown to be true using any kind of sound reasoning.

If this sentence is false, it can be shown to be true using sound reasoning, but sound reasoning cannot establish the truth of a false sentence. So sound reasoning leads to the conclusion that the sentence is true—but then it is false. Perhaps we should describe it, rather, as meaningless? But then, surely, it cannot be shown to be true using any kind of sound reasoning, so. . . .

A popular variant of this argument seeks to establish the existence of a "Gödel sentence" for any particular person, one that can unproblematically be held to be true by everybody except that person: for example,

> John will never be able to convince himself of the truth of this sentence.

And of course, in the version using "cannot be shown to be true," we are approaching the classical paradox of the Liar, which leads to the conclusion that the Liar sentence

> This sentence is false

is true if and only if it is false.

The many arguments and ideas surrounding the Liar sentence and the various "Gödel sentences" formulated above will not be discussed in this book. The following observations are however relevant. The incompleteness theorem is a mathematical theorem precisely because the relevant notions of truth and provability are mathematically definable. Nonmathematical "Gödel sentences" and Liar sentences give rise to prolonged (or endless) discussions of just what is meant by a proof, by a true statement, by sound reasoning, by showing something to be true, by convincing oneself of something, by believing something, by a meaningful statement, and so on. In spite of the similarities in form between these other sentences and the fixpoints of arithmetically (or more generally, mathematically) definable properties of sentences in a formal language, we are again not dealing with any application of the incompleteness theorem or its proof, but with considerations or conundrums inspired by the incompleteness theorem. It is an open question whether the pleasant confusion that statements like "John will never accept this statement as true" tend to create in people's minds has any theoretical or philosophical significance.

4.4 Incompleteness and the TOE

The TOE is the hypothetical Theory of Everything, which is sometimes thought to be an ideal or Holy Grail of theoretical physics. The incompleteness theorem has been invoked in support of the view that there is no such theory of everything to be had, for example, by eminent physicists Freeman Dyson and Stephen Hawking.

In a book review in the New York Review of Books, Dyson writes:

> Another reason why I believe science to be inexhaustible is Gödel's theorem. The mathematician Kurt Gödel discovered and proved the theorem in 1931. The theorem says that given any finite set of rules for doing mathematics, there are undecidable statements, mathematical statements that cannot either be proved or disproved by using these rules. Gödel gave examples of undecidable statements that cannot be proved true or false using the normal rules of logic and arithmetic. His theorem implies that pure mathematics is inexhaustible. No matter how many problems we solve, there will always be other problems that cannot be solved within the existing rules. Now I claim that because of Gödel's theorem, physics is inexhaustible too. The laws of physics are a finite set of rules, and include the rules for doing mathematics, so that Gödel's theorem applies to them. The theorem implies that even within the domain of the basic equations of physics, our knowledge will always be incomplete.

It seems reasonable to assume that a formalization of theoretical physics, if such a theory can be produced, would be subject to the incompleteness theorem by incorporating an arithmetical component. However, as emphasized in Section 2.3, Gödel's theorem only tells us that there is an incompleteness in the arithmetical component of the theory. The basic equations of physics, whatever they may be, cannot indeed decide every arithmetical statement, but whether or not they are complete considered as a description of the physical world, and what completeness might mean in such a case, is not something that the incompleteness theorem tells us anything about.[1]

Another invocation of incompleteness goes further:

> Not to mention there are an infinite number of other attributes of the world which are simply not quantifiable or computable, such as beauty and ugliness, happiness and misery, intuition and inspiration, compassion and love etc. These are completely outside the grasp of any mathematical Theory of Everything. Since scientific theories are built upon mathematical systems, incompleteness must be inherited in all our scientific knowledge as well. The incompleteness theorem reveals that no matter what progress is made in our science, science can never in principle completely disclose Nature.

Here the connection with the actual content of the incompleteness theorem is tenuous in the extreme: "Since scientific theories are built upon mathematical system, incompleteness must be inherited in all our scientific knowledge as well." This doesn't follow, since nothing in the incompleteness theorem excludes the possibility of our producing a complete theory of stars, ghosts, and cats all rolled into one, as long as what we say about stars, ghosts, and cats cannot be interpreted as statements about the natural numbers. That science cannot be expected to disclose to us everything about beauty and ugliness, intuition and inspiration, and so on, is a reasonable view which neither needs nor is supported by Gödel's theorem.

Stephen Hawking, in a talk entitled "Gödel and the End of Physics," also mentions Gödel's theorem:

> What is the relation between Gödel's theorem, and whether we can formulate the theory of the universe, in terms of a finite number of principles? One connection is obvious. According to the positivist philosophy of science, a physical theory is a mathematical model. So if there are mathematical results that cannot be proved, there are physical problems that cannot be pre-

[1] Dyson conceded this point in a gracious response to similar remarks made by Solomon Feferman in a letter to the New York Review of Books (July 15, 2004).

> dicted. One example might be the Goldbach conjecture. Given an even number of wood blocks, can you always divide them into two piles, each of which cannot be arranged in a rectangle? That is, it contains a prime number of blocks. Although this is incompleteness of sorts, it is not the kind of unpredictability I mean. Given a specific number of blocks, one can determine with a finite number of trials, whether they can be divided into two primes. But I think that quantum theory and gravity together introduce a new element into the discussion, one that wasn't present with classical Newtonian theory. In the standard positivist approach to the philosophy of science, physical theories live rent-free in a Platonic heaven of ideal mathematical models. That is, a model can be arbitrarily detailed and can contain an arbitrary amount of information, without affecting the universes they describe. But we are not angels who view the universe from the outside. Instead, we and our models are both part of the universe we are describing. Thus, a physical theory is self-referencing, like in Gödel's theorem. One might therefore expect it to be either inconsistent, or incomplete. The theories we have so far, are both inconsistent, and incomplete.

Here the upshot is that physical theory is "self-referencing," apparently in the sense that physical theories are "part of the universe" and that one might therefore expect them to be inconsistent or incomplete, considering that Gödel proved his first incompleteness theorem using a self-referential statement. Again, the relevance of the incompleteness theorem is here at most a matter of inspiration or metaphor. But Hawking also touches on another subject, the relevance of arithmetic to predictions about the outcome of physical experiments. Given $104,729$ wooden blocks, will we succeed in an attempt to arrange them into a rectangle? A computation shows $104,729$ to be a prime, so we conclude that no such attempt will succeed. Or, to take a somewhat more realistic example, consider the 15-puzzle, the still-popular sliding square puzzle that Sam Lloyd introduced in 1873, which has long been a favorite among AI researchers when testing heuristic search algorithms. Lloyd offered a $1,000 reward for the solution of the "15-14 problem," the problem of rearranging the squares so that only the last two squares were out of place. He well knew that his money was not at risk, since a combinatorial argument shows that the problem has no solution. Thus, he could set people to work on the problem and confidently predict, on the basis of arithmetical reasoning, the eventual outcome (their giving up).

Do such examples show that arithmetical incompleteness can entail an incompleteness in our description of the physical world? Not really. Suppose the Diophantine equation $D(x_1, \ldots, x_n) = 0$ has no solution, but this fact is not provable in our mathematics. We then have no basis for a prediction of the outcome of any physical experiment describable as "searching for a solution of the equation $D(x_1, \ldots, x_n) = 0$." (Such an experiment might consist in people rearranging wooden blocks or doing pen-and-paper calculations, or it might consist in having a computer execute a program.) This does not, however, indicate any incompleteness in our description of the physical systems involved. Our predictions of the outcome of physical experiments using arithmetic are based on the premise that arithmetic provides a good model for the behavior of certain actual physical systems with regard to certain observable properties (which in particular implies that physical objects like blocks of wood have a certain stability over time, that there are no macroscopic tunneling effects that render arithmetic inapplicable, that eggs do not spontaneously come into existence in baskets, and so on). The relevant description of the physical world amounts to the assumption that this premise is correct. The role of the arithmetical statement is as a premise in the application of this description to arrive at conclusions about physical systems.

4.5 Theological Applications

Gödel sometimes described himself as a theist and believed in the possibility of a "rational theology," although he did not belong to any church. In [Wang 87] he is quoted as remarking that "I believe that there is much more reason in religion, though not in the churches, that one commonly believes." It should not be supposed that Gödel's theism agreed with that expressed in established theistic religions. Theistic religions usually involve a God or several gods assumed to stand in a relationship to human beings that makes it meaningful to pray to the God(s), to thank the God(s), to obey the God(s), and more generally to communicate with the God(s). Gödel's "rational theology" was not concerned with such matters. Among his unpublished papers was a version of St. Anselm's ontological proof of the existence of God. More precisely, the conclusion of the argument is that there is a God-like individual, where x is defined to be God-like if every essential property of x is positive and x has every positive property as an essential property. As this explanation of "God-like" should make clear, Gödel's idea of a rational theology was not of an evangelical character, and Oskar Morgenstern relates ([Dawson 97, p. 237]) that he hesitated to publish the proof "for fear that a belief in God might be ascribed to him, whereas, he said, it was undertaken as a purely logical investigation, to demonstrate that such a proof could be carried out on the basis of accepted principles of formal logic."

Although Gödel was thus not at all averse to theological reasoning, he did not attempt to draw any theological conclusions from the incompleteness theorem. However, others have invoked the incompleteness theorem in theological discussions. *Bibliography of Christianity and Mathematics*, first edition 1983, lists 13 theological articles invoking Gödel's theorem. Here are some quotations from the abstracts of these articles:

> Nonstandard models and Gödel's incompleteness theorem point the way to God's freedom to change both the structure of knowing and the objects known.
>
> Uses Gödel's theorem to indicate that physicists will never be able to formulate a theory of physical reality that is final.
>
> Stresses the importance of Gödel's theorems of incompleteness toward developing a proper perspective of the human mind as more than just a logic machine.
>
> ...theologians can be comforted in their failure to systematize revealed truth because mathematicians cannot grasp all mathematical truths in their systems, either.
>
> If mathematics were an arbitrary creation of men's minds, we can still hold to eternal mathematical truth by appealing to Gödel's incompleteness result to guarantee truths that can be discovered only by the use of reason and not by the mechanical manipulation of fixed rules—truths which imply the existence of God.
>
> It is argued by analogy from Gödel's theorem that the methodologies, tactics, and presuppositions of science cannot be based entirely upon science; in order to decide on their validity, resources from outside science must be used.

As can be seen from these quotations, appeals to the incompleteness theorem in theological contexts are sometimes invocations of Gödelian arguments in the philosophy of mind, which will be considered in Chapter 6, and sometimes follow the same line of thought as the arguments considered in Section 4.4 in connection with "theories of everything." But there are also some more specifically theological appeals to the theorem. These are sometimes baffling:

> For thousands of years people equated consistency with determinism, holding that a logically consistent sequence of propositions could have only one outcome. This feeling lies behind the notion that God knows and controls everything. Kurt Gödel, working on a question asked by David Hilbert, showed that consistency does not always mean determinism.

It is difficult to know what to make of the idea that "a logically consistent sequence of propositions can have only one outcome" or of its relation to negation completeness.

Other theological invocations of Gödel are more easily made sense of. The following reflections by Daniel Graves are taken from an essay on the "Revolution against evolution" website:

> Gödel showed that "it is impossible to establish the internal logical consistency of a very large class of deductive systems—elementary arithmetic, for example—unless one adopts principles of reasoning so complex that their internal consistency is as open to doubt as that of the systems themselves." (Here the author is quoting [Nagel and Newman 59]). In short, we can have no certitude that our most cherished systems of math are free from internal contradiction.
>
> Take note! He did not prove a contradictory statement, that A = non-A, (the kind of thinking that occurs in many Eastern religions). Instead, he showed that no system can decide between a certain A and non-A, even where A is known to be true. Any finite system with sufficient power to support a full number theory cannot be self-contained.
>
> Judeo-Christianity has long held that truth is above mere reason. Spiritual truth, we are taught, can be apprehended only by the spirit. This, too, is as it should be. The Gödelian picture fits what Christians believe about the universe. Had he

> been able to show that self-proof was possible, we would be in deep trouble. As noted above, the universe could then be self-explanatory.
>
> As it stands, the very real infinities and paradoxes of nature demand something higher, different in kind, more powerful, to explain them just as every logic set needs a higher logic set to prove and explain elements within it.
>
> This lesson from Gödel's proof is one reason I believe that no finite system, even one as vast as the universe, can ultimately satisfy the questions it raises.

A main component of these reflections is the observation that any consistent system incorporating arithmetic cannot prove itself consistent and cannot answer every question it raises, and is in these respects "not self-contained." It is only in a wider, stronger system that every question raised by the first system can be answered and more besides, and the first system can also be proved consistent. As a description of the incompleteness theorem, this is unobjectionable, and in fact in his 1931 paper Gödel had a footnote 48a that should be quite congenial to the author of passage quoted:

> As will be shown in Part II of this paper, the true reason for the incompleteness inherent in all formal systems of mathematics is that the formation of ever higher types can be continued into the transfinite (see *Hilbert 1926*, page 184), while in any formal system at most denumerably many of them are available. For it can be shown that the undecidable propositions constructed here become decidable whenever appropriate higher types are added (for example, the type ω to the system P). An analogous situation prevails for the axiom system of set theory.

While this part of Graves' comments is thus unobjectionable, they are highly dubious in other aspects. The idea that the consistency of arithmetic is in doubt will be commented on in Chapter 6. It is not clear that this idea is theologically relevant. The formulation "no system can decide between a certain A and non-A" is incorrect, since there is no A such that no system can decide between A and non-A, but rather for any given system there is an A depending on the system such that the system cannot decide between A and non-A. But the main thrust of the passage is (unsurprisingly) to point to the incompleteness theorem as providing an analogy to a Christian perception of the relation between the universe and God: the universe needs something higher to explain it. Finding such analogies is perfectly legitimate, but of course it is tendentious in the extreme to speak of analogies as "lessons." It is also obscure on what grounds the author claims that "Had he been able to show that self-proof was possible, we would be in deep trouble." If "self-proof" means a consistency proof for S carried out within S itself, it is difficult to believe that any theologian would have concluded from a consistency proof for PA carried out within PA that since arithmetic needs nothing higher than itself to support or explain it, neither does the universe, and so God is unnecessary.

The author makes a further comment:

> As a third implication of Gödel's theorem, faith is shown to be (ultimately) the only possible response to reality. Michael Guillen has spelled out this implication: "the only possible way of avowing an unprovable truth, mathematical or otherwise, is to accept it as an article of faith." In other words, scientists are as subject to belief as non-scientists. [The reference is to *Bridges to Infinity*, Los Angeles: Tarcher, 1983, p.117.]

Here, the reference to Gödel's theorem is pointless. That science involves faith is a standard argument in discussions of theology and religion, but one to which Gödel's theorem is irrelevant. As much or as little faith is needed to accept the axioms of a theory whether or not that theory is complete and the necessity of accepting some basic principles without proof is not something that was revealed by Gödel's theorem.

Another example of a theological invocation of Gödel's theorem is given by the following comments by Najamuddin Mohammed:

> It is pointed out that, no matter how you describe the world (with logical rules) there will always be "some things" that you cannot determine as true or false. And whether you select the answer to these "some things" as true or false doesn't affect the validity of your logical rules. Strange but true!
>
> For example let's say: you and I have agreed upon a set of logical rules, then there will always be some thing, lets call it A, that we cannot determine as true or false, using our logical rules. You can take A to be true and I can take A to be false, but in either case we are both logically consistent with our new set of logical rules respectively.

> But now we have two sets of self-consistent rules and again there will always be something called B that we cannot agree upon....and so on. This is the basis of Gödel's Incompleteness Theorem.
>
> If we rely on logic or reason alone we can end up in utter confusion, with many contradictory but logically self-consistent systems of reasoning/logic. Which is correct? Does everything depend on our current psychological disposition as to what is right and wrong? Correctness has no meaning in these cases, all this can lead to agnostic and atheistic stances.

There is a similarity between these reflections and the ideas about a "postmodern condition" created by the incompleteness theorem commented on in Section 2.8. Incompleteness, it is argued, leads to a profusion of different consistent theories, and nobody knows where truth—or "truth"—is to be had. Thus, reason alone cannot put us on the right path, and religious faith is the way to go. Again, the weakness of this line of thought is that there is not in fact any such branching off into various directions in mathematical thinking, no floundering in a sea of undecidability. The "utter confusion" in mathematical thinking is a theological dream only.

What remains to be considered in connection with theological invocations of the incompleteness theorem are two lines of thought that are not specifically theological but are often thought congenial from a theological point of view: the skeptical conclusions thought to follow from the second incompleteness theorem, and the conclusions about the nature of the human mind claimed to follow from the first incompleteness theorem. These will be considered in later chapters.

5

Skepticism and Confidence

5.1 The Second Incompleteness Theorem

The Discovery of the Theorem

Gödel first presented his incompleteness theorem at a conference on "Epistemology of the exact sciences" in 1930. Gödel was then not yet 25 years old, so he is one of a considerable number of mathematicians and logicians who have made major discoveries at an early age. Among those present was the Hungarian mathematician John von Neumann, three years Gödel's senior, one of the great names of twentieth-century mathematics and the subject of various anecdotes about his remarkable powers of quick apprehension. It appears that he was the one participant at the conference who immediately understood Gödel's proof. At this point Gödel had not yet arrived at his second incompleteness theorem, and his proof of the first incompleteness theorem was not applicable to PA, but only to somewhat stronger theories. His proof did, however, establish that assuming a theory S to which the proof applied to be consistent, it follows that the Gödel sentence G for S is unprovable in S. Reflecting on this after the conference, von Neumann realized that the argument establishing the implication "if S is consistent, then G is not provable in S" can be carried out within S itself. But then, since G is equivalent in S to "G is not provable in S," it follows that if S proves the statement Con_S expressing "S is consistent" in the language of S, S proves G, and hence is in fact inconsistent. Thus, the second incompleteness theorem follows: if S is consistent, Con_S is not provable in S. By the time von Neumann had discovered this and written to Gödel about it, Gödel had himself already made the same discovery and included it in his recently accepted 1931 paper.

Central to this proof of the second incompleteness theorem is the notion of an ordinary mathematical proof being *formalizable* in a certain formal system. This means that for every step in the proof there is a corresponding series of applications of formal rules of inference in the system, so that the conclusion of the proof, when expressed in the language of the system, is also a theorem of the system. At the time of Gödel's proof, this notion was familiar to logicians and philosophers of mathematics through the work of Gottlob Frege, Russell and Whitehead, Hilbert, and others, so Gödel could take it for granted in arguing that the proof of the first half of the incompleteness theorem for P—"if P is consistent then G is not provable in P"—was formalizable in P itself.

In fact, Gödel only sketched the proof of the second incompleteness theorem in his paper. To prove the second (as opposed to the first) incompleteness theorem for a formal system S, we definitely need to arithmetize the syntax of S, and reason about S in S itself, since this is required to even express "S is consistent" in the language of S. This arithmetization was carried out in detail in Gödel's paper, but we then need to verify that the proof of the implication "if S is consistent then G is unprovable in S" is indeed formalizable in S. In his paper, Gödel only presented this as a plausible claim, noting that the proof of the first incompleteness theorem only used elementary arithmetical reasoning of a kind formalizable in the system P for which he carried out his proof. In the planned follow-up to his paper, he intended to give a full proof of the second incompleteness theorem. Part II of the paper never appeared, for the informal argument Gödel gave was in fact quite convincing to his readers, and furthermore, in 1939, in the two-volume work *Grundlagen der Mathematik* (Foundations of Mathematics) by Paul Bernays and David Hilbert, a detailed proof was given. At that point, the general concept of a formal system had been clarified through the work of Turing and Church, yielding the general formulation of the incompleteness theorem that we know today.

For the proof of the second incompleteness theorem, what was needed was to show that the implication "if Con_S then G" is provable in S. The converse implication, "if G then Con_S," is much more easily shown to be provable in S. All that is needed is to formalize in S the argument that G implies that G is not provable in S, and so it also implies that S is consistent, since everything is provable in an inconsistent system. Thus, G

and Con_S are *equivalent*, and this equivalence is provable in S. It follows that we know the Gödel sentence G for a formal system to be true if and only if we know the system to be consistent.

That G and Con_S are equivalent in S merits underlining, for two reasons. First, it is often said that Gödel's proof shows G to be true, or to be "in some sense" true. But the proof does not show G to be true. What we learn from the proof is that G is true if and only if S is consistent. In this observation, there is no reason to use any such formulation as "in some sense true"—if S is consistent, G is true in the ordinary mathematical sense of "true," as when we say that Goldbach's conjecture is true if and only if every even number greater than 2 is the sum of two primes. Second, Gödel's proof does not show that there is any arithmetical statement at all that we know to be true but is not provable in S, since the statement "Con_S if and only if G" is provable in S itself, for the theories S here at issue. For many theories S we do know that S is consistent, and hence know G to be true, but there is nothing in Gödel's proof that shows S to be consistent.

Another aspect of the second incompleteness theorem that needs to be emphasized is that it does not show that S can only be proved consistent in a system that is *stronger* than S. To say that S is consistent is not to endorse the axioms of S, and a proof that S is consistent need not use the axioms of S itself. An example worth keeping in mind is that by the second incompleteness theorem, PA + not-$\mathrm{Con}_{\mathrm{PA}}$ is consistent, given that PA is. It would be unfortunate if we could only prove the consistency of this theory in a stronger theory, one that in its arithmetical component proves every theorem of PA + not-$\mathrm{Con}_{\mathrm{PA}}$. Instead, we prove PA + not-$\mathrm{Con}_{\mathrm{PA}}$ consistent by proving PA consistent, and thus prove the consistency of this theory in an *incompatible* theory.

Here it should be noted that in logic, PA + Con_S is sometimes said to be stronger than S "in the sense of interpretability." The completeness theorem for first-order logic (see Chapter 7) implies that every consistent first-order theory S has a *model*, a mathematical structure in which all the axioms of S are true. A refinement of this theorem shows that such a model can in fact be defined in PA and the axioms of S can be proved to hold in PA + Con_S when interpreted in terms of this model. Thus, for example, we can define in PA a certain arithmetical relation such that when we interpret the symbol for set membership in a sentence of ZFC as referring to this relation, A' is provable in PA + $\mathrm{Con}_{\mathrm{ZFC}}$ for every theorem A of ZFC, where A' is the resulting arithmetical interpretation of A.

This does not mean that every theorem in the arithmetical component of ZFC is provable in PA + $\mathrm{Con}_{\mathrm{ZFC}}$. Suppose, for example, that the formalization A in the language of ZFC of the twin prime conjecture (a certain statement about finite sets, which we know to be true if and only if the twin prime conjecture is true) is provable in ZFC. Its translation A' is then provable in PA + $\mathrm{Con}_{\mathrm{ZFC}}$. There is no guarantee, however, that A' is provably equivalent in PA to the twin prime conjecture itself. The translation A' does not preserve the arithmetical meaning of A. Indeed, there are statements A in the arithmetical component of ZFC such that A is true but A' false (such A are undecidable in ZFC). However, if A is a *Goldbach-like* statement in the arithmetical component of ZFC, not only A', but the arithmetical statement whose meaning A captures, is provable in PA + $\mathrm{Con}_{\mathrm{ZFC}}$.

The proof of the second incompleteness theorem in the *Grundlagen* did not answer every question about how to formulate and prove the theorem in complete generality, and it was only in 1960 that Solomon Feferman cleared up the remaining areas of uncertainty (in his paper "Arithmetization of Metamathematics in a General Setting"). However, in the case of the theories actually used and studied in logic and mathematics, like PA or ZFC, it has been clear since the 1930s how to formulate and prove the second incompleteness theorem.

Gödel's original proof of the second incompleteness theorem is still the most important, in terms of the insight it gives into the theorem and its range of application, but there are several other proofs of the theorem for specific theories, again typically PA and ZFC, in the logical literature. These other proofs, however, are highly technical.

Some Consequences

The consistency of any theory T is of course provable in other theories, in the sense that these other theories have among their theorems the statement "T is consistent" (which may or may not be true). For example, if we add the axiom Con_T to PA we get a theory in which the consistency of T is provable. To reiterate, if T is consistent, PA + Con_T is stronger than T in one respect, since it proves an arithmetical theorem not provable in T, but at the same time T may prove, even in its arithmetical component, any number of (non-Goldbach-like) theorems not provable in PA.

The observation that the consistency of T is provable in PA + Con_T is not as pointless as it may seem, since it serves to underline an important distinction. The second incompleteness theorem is a theorem about *formal provability* (which is always relative to some formal system), showing that Con_T is not (for the T at issue) provable in T itself. That is, it shows that there is no formal derivation of Con_T in the theory T. It does not tell us whether "T is consistent" can be proved in the sense of being shown to be true by a conclusive argument, or by an argument acceptable to mathematicians. Neither the provability of Con_T in PA + Con_T nor the unprovability of Con_T in T itself has any immediate implications for the question whether it is possible to demonstrate the truth of Con_T to the satisfaction of the mathematical community.

Another aspect of this distinction deserves a comment. Since it is very easy to find a theory in which the consistency of T is formally provable, to say that Con_T *cannot* be formally proved in T itself is to make a stronger statement than one might naturally suppose. It is not just a matter of T lacking the means to analyze or justify itself, but one of T not being able to consistently *assert* its own consistency. Using Gödel's fixpoint construction described in Section 2.7, we can produce an arithmetical statement C which formalizes "this sentence is consistent with the axioms of PA." Thus, the theory PA + C postulates its own consistency (without attempting any analysis or justification), and so by the second incompleteness theorem, it follows that PA + C is in fact inconsistent. The unprovability of consistency might thus just as well be called the *unassertibility* of consistency. A consistent theory T cannot postulate its own consistency, although the consistency of T can be postulated in another consistent theory. (An exercise for the reader: Is it possible to have a pair of consistent theories S and T such that each postulates the consistency of the other? It follows from what has been said above that the answer is no.)

Using the second incompleteness theorem to show that a theory is inconsistent since it proves its own consistency is not uncommon in logic, although the interest of such proofs usually lies in some positive aspect of the argument. An illustration of this is given by *Löb's theorem*. Suppose we produce a provable fixpoint for the property of *being* a theorem of PA, instead of (as in Gödel's proof) the property of *not* being theorem of PA. In other words, let H (for Leon Henkin, who first raised this question) be an arithmetical statement formalizing "this statement is provable in PA." Is H provable in PA or not? Reflecting on the meaning of H only yields that H is true if and only if it is provable in PA, which doesn't get us anywhere, and it may at first seem hopeless to decide whether this strange self-referential sentence is provable in PA or not.

Löb solved the question of the provability of H by showing that it holds more generally that if PA proves "if PA proves A then A," then PA proves A (so that in particular the Henkin sentence is indeed provable in PA). To see this, consider the theory PA + not-A. We are assuming that PA proves "if PA proves A then A," which implies "if not-A then PA does not prove A." But this in turn implies "if not-A then PA + not-A is consistent," so PA + not-A proves the consistency of PA + not-A, and thus is in fact inconsistent, implying that PA proves A. The same argument applies to any theory T for which the second incompleteness theorem holds.

The second incompleteness theorem is a special case of Löb's theorem, as we see by choosing A in the formulation of the theorem as a logical contradiction "B and not-B." For such an A, "PA proves A" is the same as saying "PA is inconsistent," and any hypothetical statement "if C then A" is logically equivalent to not-C. Thus, Löb's theorem tells us that if PA proves "PA is not inconsistent" then PA is inconsistent, or in other words, if PA is consistent then PA does not prove "PA is consistent."

Löb's theorem gives us a rather odd principle for proving theorems about the natural numbers. In order to prove A, it is admissible to assume as a premise that A is provable in PA, as long as the argument is one that can be formalized in PA. For if there is a proof in PA of "if there is a proof in PA of A then A," then there is a proof in PA of A. A similar principle holds, for example, for ZFC. This principle doesn't have any known application in proving any ordinary mathematical theorems about primes or other matters of traditional mathematical interest, but it does have uses in logic.

Löb's principle may seem baffling rather than just odd. How can it be admissible, in proving A, to assume that A is provable in PA? After all, whatever is provable in PA is true, so can't we then conclude without further ado that A is true, and so prove A without doing any reasoning at all? The essential point here is that it is admissible to assume that A is provable in PA when proving A *only* if the reasoning leading from the assumption that A is provable in PA to the conclusion A can in fact be carried out within PA. That everything provable in PA is true is not something that can be established within PA itself, as is shown by the second incompleteness theorem. To say that, for example, $0 = 1$ is true if provable in PA is (given that 0 is not equal to 1) the same as saying that $0 = 1$ is not provable in PA, or in other words that PA is consistent.

Theories That Almost Prove Their Own Consistency

Since PA is consistent, it does not prove its own consistency. But if we choose any *finite* number of axioms of PA, the consistency of that finite set of axioms *is* provable in PA. The same is true for ZFC: if we take any finite set of axioms of ZFC, their consistency can be proved in ZFC.

If a theory is inconsistent, there is a proof of a logical contradiction from the axioms of the theory. For theories such as PA and ZFC, which formalize proofs that we actually carry out in mathematics, a proof is a finite sequence and can use only finitely many axioms of the theory. So if every finite subset of the axioms of the theory is consistent, then so is the whole theory. This easy argument can be carried out within PA, so it is provable in PA that if every finite subset of the axioms of PA is consistent, then so is PA. So if PA proves every finite set of PA-axioms to be consistent, then why does PA not prove itself consistent?

The answer lies in an ambiguity in the formulation "PA proves every finite set of PA-axioms to be consistent." What is true is that for any finite set M of axioms of PA, PA proves "the theory with axioms M is consistent." It is not the case however that PA proves the statement "for any finite set M of axioms of PA, the theory with axioms M is consistent." This distinction, which does not correspond to anything in ordinary informal talk about proofs and consistency, is a reminder that there are some subtleties connected with Gödel's theorem that may or may not be relevant to the many philosophical or informal ideas associated with the theorem. We will return to this point in Section 6.4, about "understanding one's own mind."

5.2 Skepticism

The incompleteness theorem is often thought to support some form of skepticism with regard to mathematics. It is argued either that we cannot, strictly speaking, prove anything in mathematics or that the consistency of theories like PA or ZF is shown to be doubtful by the theorem.

In many cases no explanation is given of how the skeptical conclusion is supposed to follow. Thus, the Encyclopedia Britannica says mysteriously of Gödel's proof that it

> ...states that within any rigidly logical mathematical system there are propositions (or questions) that cannot be proved or

> disproved on the basis of the axioms within that system and that, therefore, it is uncertain that the basic axioms of arithmetic will not give rise to contradictions.

Other peculiar aspects of this comment aside (it is actually borrowed from Carl Boyer's and Uta Merzbach's *History of Mathematics*), the abrupt conclusion from incompleteness to possible inconsistency has no immediately apparent rational basis, and none is suggested in the article. Similarly, we find such comments as

> By Gödel's theorem, a system is either incomplete or inconsistent. Thus, logically speaking, it is impossible for us to fully "prove" any proposition.

The occurrence in such remarks of phrases like "logically speaking" is a noteworthy feature of many startling *non sequiturs* inspired by the incompleteness theorem.

When accompanied by an intelligible argument, skeptical conclusions based on Gödel's work usually specifically invoke the second incompleteness theorem. Thus, Nagel and Newman [Nagel and Newman 59, p. 6] state that Gödel proves

> that it is impossible to establish the internal logical consistency of a very large class of deductive systems—elementary arithmetic, for example—unless one adopts principles of reasoning so complex that their internal consistency is as open to doubt as that of the systems themselves.

Other commentators go further. Morris Kline, in *Mathematics: The Loss of Certainty*, states that "Gödel's result on consistency says that we cannot prove consistency in any approach to mathematics by safe logical principles."

There are two main ingredients in such reflections: the idea that the consistency of some or all of the formal systems we use in mathematics is *doubtful*, and the idea that the consistency of these systems cannot be *proved* in the same sense as other mathematical statements can be proved. For a perspective on these ideas, let us begin with the matter of doubt.

The Irrelevance of Gödel's Theorem to Doubts

Nothing in Gödel's theorem in any way contradicts the view that there is no doubt whatever about the consistency of any of the formal systems that we use in mathematics. Indeed, nothing in Gödel's theorem is in any way incompatible with the claim that we have absolutely certain knowledge of the *truth* of the axioms of these systems, and therewith of their consistency.

In considering this point, we need to distinguish between two things: what degree of skepticism or confidence regarding mathematical axioms or methods of reasoning is justifiable or reasonable, and what bearing Gödel's theorem has on the matter. Perhaps we take a dim view of the claim that we know with absolute certainty the truth of, say, the axioms of ZFC, but how can we use Gödel's theorem to criticize this claim? Can we direct at the claim the telling criticism that if we know with absolute certainty that the axioms of ZFC are true, then the consistency of ZFC must be provable in ZFC itself? No, because this is not a telling criticism at all. Why should there be a proof of the consistency of ZFC in ZFC just because we know with absolute certainty that the axioms of ZFC are true (and hence consistent)? Obviously, we cannot prove everything in mathematics. We don't need Gödel's theorem to tell us that we must adopt some basic principles without proof. And given that the axioms of ZFC are so utterly compelling, so obviously true in the world of sets, we can do no better than adopt these axioms as our starting point. Since the axioms are true, they are also consistent.

Again, the point at issue is not whether such a view of the axioms of ZFC is justified, but whether it makes good sense to appeal to the incompleteness theorem in criticism of it. *If* the axioms of ZFC are manifestly true, they are obviously consistent, although there is no reason to expect a consistency proof for ZFC in ZFC.

From the point of view of a skeptic about the consistency of ZFC, it is on closer inspection also unclear what is supposed to be the relevance of the second incompleteness theorem. What would be the interest of a consistency proof for ZFC given in ZFC? Since the consistency of ZFC is precisely what is in question, there is no reason to expect such a proof to carry any weight.

So if we have no doubts about the consistency of ZFC, there is nothing in the second incompleteness theorem to give rise to any such doubts. And if we do have doubts about the consistency of ZFC, we have no reason to believe that a consistency proof for ZFC formalizable in ZFC would do anything to remove those doubts.

The Tradition of Finitism

Hilbert had the idea of proving the consistency of strong theories like ZFC on the basis of very weak mathematical assumptions and finitistic reasoning, without assuming the existence of infinite sets and making only restricted use of logical principles. The second incompleteness theorem does indeed establish that we cannot prove the consistency even of PA using only the kind of reasoning that Hilbert had in mind. So if we take the view that only finitistic reasoning in Hilbert's sense embodies "safe logical principles" and is not open to doubt as regards its consistency, or that only finitistic proofs are really proofs, we will indeed conclude that the consistency of even elementary arithmetic cannot be proved by safe logical principles, and so on. But such a view is in no way a *conclusion* from Gödel's theorem. It is a particular doctrine in the philosophy of mathematics that one brings to Gödel's theorem. Those who do not believe that only finitistic reasoning is unproblematically correct or meaningful can accept with equanimity that there is no finitistic consistency proof for PA and observe that the consistency of PA is easily provable by other means.

It should also be noted that from a less narrow viewpoint than that of finitism, consistency is only a weak soundness condition. That S is consistent does not, as we know from the second incompleteness theorem itself, rule out that S proves false theorems. For example, PA + not-Con_{PA} is consistent but falsely proves the inconsistency of PA (and thus of itself). If we wish to justify our theories, a mere consistency proof will not take us far.

Not only philosophers, but also mathematicians, not infrequently seem to get carried away by the philosophical legacy of Hilbert and the decades of rhetoric surrounding the incompleteness theorem and, without explicitly endorsing any finitistic doctrines, attach a large significance to the impossibility of giving a finitistic consistency proof for PA. A consistency proof, they say, can only be a *relative* consistency proof, showing, for example, that if ZFC is consistent, then PA is consistent. We need next to take a closer look at the idea that consistency proofs are somehow not just ordinary mathematical proofs, or that consistency statements cannot be proved in the sense that other mathematical statements can be proved.

5.3 Consistency Proofs

There are many consistency proofs in the logical literature. Georg Kreisel has made the point that these proofs often prove something a great deal more interesting than mere consistency, but putting one's finger on just what that something is can be difficult. An example of this is the consistency proof for PA that was given by Gerhard Gentzen in 1935. This proof used an induction principle for an ordering relation much "longer" than that of the natural number series, but applied that induction principle only to a restricted class of properties. Thus, it extended PA in one direction but restricted it in another. Gentzen's proof, along with the whole subject of "ordinal analysis" to which it gave rise, is very technical and stands in no simple relation to any doubts that people may have about the consistency of PA.

Another famous example is Gödel's proof of the "relative consistency" of the axiom of choice in set theory. This is not a consistency proof, but is usually described as a proof of the implication "if ZF is consistent then ZFC is consistent." In this formulation, Gödel's result conveys no information of interest to anybody who regards ZFC as obviously consistent. But, in fact, the proof establishes much more and shows, for example, that every *arithmetical* theorem that can be proved using the axiom of choice can be proved without using that axiom—a fact that is by no means obvious, even given the consistency of ZFC.

These two famous results from logic have been mentioned here chiefly in order to emphasize that the content and interest of "consistency proofs" or "relative consistency proofs" in mathematical logic is often a technical and difficult matter. The remaining comments will concern only consistency proofs of a different kind. These proofs are mathematically essentially trivial, and they prove much more than the consistency of a theory, namely that *all* theorems of the theory are true.

Proving ZFC Consistent

In the case of ZFC, there is no technical consistency proof in logic corresponding to Gentzen's consistency proof for PA. The consistency of ZFC is, however, an immediate consequence of various much stronger statements than "ZFC is consistent" and follows in particular from set-theoretical principles of a kind known as *axioms of infinity* (see Section 8.3). Because such axioms (with the exception of the basic axiom of infinity which is part of

ZFC) are not ordinarily used in mathematics to prove theorems, it is a perfectly reasonable observation that these consistency proofs for ZFC are not ordinary mathematical proofs, and that such proofs do not establish that the consistency of ZFC can be proved in the same sense as Fermat's Last Theorem has been proved. This is so whether or not we regard axioms of infinity as mathematically justified. As long as mathematicians, with very few exceptions, do not make use of axioms of infinity in proving theorems, it makes sense to observe that there is no "ordinary mathematical proof" of the consistency of ZFC.

Similar comments apply to what is known as proofs by *reflection* of the consistency of ZFC. As has been noted, for every finite set M of axioms of ZFC, it is provable in ZFC that M is consistent. So, it may seem, if we are prepared in mathematics to accept every statement that we know to be a theorem of ZFC as proved, we should accept it as proved that every finite set of axioms of ZFC is consistent, and thereby that ZFC is consistent. To examine this line of argument would be to open a philosophical can of worms that had better be set aside in this context. So we will just note again that whatever the merits of such a consistency proof by reflection, it does not much resemble proofs in ordinary mathematics.

On the other hand, it is a simple matter to prove PA consistent using only ordinary mathematics. So in order to take a closer look at the idea that consistency proofs as such are necessarily more problematic, more dubious, or less robust than other mathematical proofs, we turn to the case of arithmetic.

Proving PA Sound

As an example of a trivial consistency proof that falls well within ordinary mathematics, consider the question of how to prove the consistency of the theories in the sequence

$$\mathrm{PA}, \mathrm{PA}_1, \mathrm{PA}_2, \ldots$$

where PA_1 is obtained by adding the axiom "PA is consistent" to PA, PA_2 adds the axiom "PA_1 is consistent" to PA_1, and so on.

We say that a theory T is *arithmetically sound* if every arithmetical theorem of T is true. An arithmetically sound theory is consistent, since an inconsistent theory proves the false arithmetical statement $0 = 1$. Furthermore, if T is arithmetically sound, the theory $T+$ "T is consistent" is also arithmetically sound. Given that PA is arithmetically sound, we find that all of the theories in the above sequence are sound, and hence consistent. And, in fact, very much longer sequences of theories can be shown to be consistent by the same reasoning. (For some hints about what can be meant by "longer sequences" here, see Section 5.4.)

Note that the knowledge that PA is *consistent* is not enough to justify the consistency of the theories in the sequence after the first, since we know from the incompleteness theorem that there are consistent theories that prove their own inconsistency. Thus, for example, the theory T obtained by adding to PA the axiom "PA is inconsistent" is consistent, but if we add to T the axiom "T is consistent" we get an inconsistent theory.

So to prove the theories in the sequence to be consistent, it is enough to prove that PA is arithmetically sound. What does this involve? Well, we need to define "true arithmetical sentence," then we need to show that the axioms of PA are all true arithmetical sentences and that the rules of reasoning of PA lead from true premises to true conclusions. Inevitably, this involves some logical and mathematical formalities, and so falls outside the treatment in this book. We will note the following aspects of the proof.

First, recall that the concept of "true arithmetical sentence" is not defined relative to any formal system. Instead, as explained in Section 2.4, what we get from a mathematical definition of the concept of truth for arithmetical sentences is simply that Goldbach's conjecture is true if and only if every even number greater than 2 is the sum of two primes, and similarly for other arithmetical sentences.

Second, the proof that the axioms of PA are true and the rules of reasoning of PA lead from true statements to true statements uses just the same axioms and rules of reasoning as those embodied in PA, plus a little bit of set theory or some mathematics at a comparable degree of abstraction. The proof is sometimes said to be carried out in ZFC, but logically speaking this is enormous overkill. Only a very much weaker set theory is needed to carry out the proof, specifically a fragment of ZFC known as ACA. Although mathematicians in general have no reason to be at all familiar with the formulation of ACA, the methods of reasoning formalized in ACA are commonly used in mathematics.

So what we have here is a mathematical proof, formalizable in the weak set theory ACA, that all the theories in the sequence PA, PA_1,... are consistent. In particular, we have a mathematical proof, using ordinary mathematical principles, that PA is consistent.

A common objection to this description of what has been achieved is that the proof is really no proof of consistency, for ACA is logically stronger than all of the theories in the sequence, and if we have doubts about the consistency of PA or any of the other theories in the sequence, these doubts will extend to the consistency of ACA. So all we can say that the proof shows is the consistency of the theories in the sequence *assuming* the consistency of ACA. In other words, the proof shows that *if* ACA is consistent, then PA and the other theories in the sequence are consistent.

This whole line of thought is predicated on the assumption that we have doubts about the consistency of PA and are trying to allay those doubts by means of a consistency proof. But when we regard the axioms and principles formalized in PA and ACA as straightforwardly part of our mathematical knowledge, the soundness proof for PA (and the other theories in the sequence) is not intended to allay any doubts at all. It is quite simply an essentially trivial proof of a basic result in logic.

There is, therefore, no basis in Gödel's theorem for the idea that a consistency proof—in this case for PA and the other theories in the sequence—is not a proof in exactly the same sense as any other mathematical proof is a proof. Every mathematical proof is based on certain basic axioms and rules of reasoning. A consistency proof such as the one sketched by no means yields a *justification* of the axioms and rules of reasoning formalized in PA. It is just a proof of an arithmetical statement, a proof which establishes the statement as true in the same way and in the same sense of "establish" as do other proofs of arithmetical statements using those same axioms and rules of reasoning. In regarding the proof as establishing the consistency of PA, we are of course drawing on our confidence in the mathematical axioms and rules of reasoning formalized in ACA—not just confidence in their consistency, but in their mathematical correctness (which might mean, in this context, their arithmetical soundness).

Those who do not regard PA and ACA as straightforwardly formalizing a part of our mathematical knowledge may, of course, have every reason to doubt any and all results proved in ACA, including the consistency proof for PA. So let us take a closer look at this skeptical perspective.

The Skeptical Perspective

If we do have doubts about the consistency of a theory T and seek to allay those doubts by means of a consistency proof for T in a theory S, then indeed we need to have confidence in the reasoning used in that particular proof in S. And if all we know is that there is *some* proof in S of the consistency of T, we need to have confidence in the consistency of S in order to conclude that T is consistent.

Such a skeptical perspective is not peculiar to those who have or claim to have doubts about the consistency of ACA or PA. Whatever our view of PA, ACA, ZFC, and other theories, there will be cases where we by no means regard the mathematical principles formalized in a theory T as evident or acceptable and have doubts even about the consistency of T. For example, the theory obtained by adding the axiom "PA is inconsistent" to PA has a false axiom, and without knowledge of Gödel's theorem it is natural to have doubts about its consistency. But as we know, the consistency of this theory can be proved, as a consequence of the consistency of PA.

What is striking about skepticism with regard to theories like PA and ACA is that it is rarely observed in any ordinary mathematical contexts. In particular, an insistence on a hypothetical interpretation of an arithmetical theorem (as in the claim that the above consistency proof only shows "if ACA is consistent then PA is consistent") is rarely heard from mathematicians or anybody else in connection with proofs of ordinary arithmetical statements. For example, one never encounters as a response to the claim "Andrew Wiles proved Fermat's Last Theorem" the objection "No, no, that's not possible—all he proved was that *if* ZFC is consistent, then there is no solution in positive integers to $x^n + y^n = z^n$ for $n > 2$." The reason for this difference is not that Wiles' proof was so elementary that the doubts affecting the consistency proof for PA do not arise. On the contrary, Wiles' proof involves heavy mathematical machinery, and it is an open question just what principles are needed to prove the theorem. But it is as true in the case of Wiles' proof as in the case of the consistency proof sketched here that *if* you have doubts about the consistency of ZFC (assuming this to be the setting for Wiles' proof), there is no obvious reason why you should accept a proof in ZFC of Fermat's Last Theorem as showing that theorem to be true.

It may be held that there is *always* an implicit "if ZFC is consistent" affixed to an arithmetical theorem proved in ZFC, and similarly for other theories. In the case of consistency theorems, it might be argued, we need to make this explicit, so as not to give the misleading impression that we have given an *absolute* consistency proof, allaying all doubts about the consistency of for example PA.

One may wonder, in such a case, why it is not equally important to avoid giving the misleading impression that Fermat's Last Theorem has been proved in any absolute sense. But there is the more substantial question of how to make sense of mathematics in general in such hypothetical terms. Consider the classical result proved in 1837 by Dirichlet, that whenever the positive integers n and k have no common divisor greater than 1, there are infinitely many primes in the sequence n, $n+k$, $n+2k$, $n+3k$,.... What can we conclude, assuming that we do not accept the mathematical principles formalized in ACA as valid, from a proof in ACA of Dirichlet's theorem? In this case, because Dirichlet's theorem is not a Goldbach-like statement, we have no grounds for concluding "if ACA is consistent, there are infinitely many primes in the sequence n, $n+k$, $n+2k$, $n+3k$,... whenever n and k have no common divisor greater than 1." So just what does the proof of Dirichlet's theorem prove?

A further consideration of these matters would take us too far into the philosophy of mathematics. The basic points argued above can be summed up as follows. It is indeed perfectly possible to have doubts about the consistency of a theory T and to seek to eliminate those doubts through a consistency proof. In such a case we need to carry out the proof in a theory whose consistency is not equally open to doubt. However, a consistency proof may just as well be a perfectly ordinary mathematical proof of a certain fact about a formal system (or, in its arithmetized form, about the natural numbers), not aiming at allaying doubts about the consistency of mathematics, any more than proofs of arithmetical theorems in general are aimed at allaying such doubts. Gödel's theorem tells us nothing about what is or is not doubtful in mathematics. To speak of the consistency of arithmetic as something that cannot be proved makes sense only given a skeptical attitude towards ordinary mathematics in general.

5.4 Inexhaustibility

Suppose we are not skeptically inclined, but rather accept some formal system to which the incompleteness theorem applies, say PA or ZFC, as unproblematically formalizing part of our mathematical knowledge. What are then the consequences of the second incompleteness theorem?

On this, Gödel commented (*Collected Works*, vol. III, p. 309, italics in the original):

> It is *this* theorem [the second incompleteness theorem] which makes the incompletability of mathematics particularly evident. For, *it makes it impossible that someone should set up a certain well-defined system of axioms and rules and consistently make*

> *the following assertion about it: All of these axioms and rules I perceive (with mathematical certitude) to be correct, and moreover I believe that they contain all of mathematics.* If somebody makes such a statement he contradicts himself. For if he perceives the axioms under consideration to be correct, he also perceives (with the same certainty) that they are consistent. Hence he has a mathematical insight not derivable from his axioms.

Thus in this case the second incompleteness theorem has the positive consequence that we can always extend any formal system that we recognize as sound (in the sense that its axioms are all true statements) to a stronger system that we also recognize as sound, by adding as a new axiom the statement that the original system is consistent.

This means that we immediately come up with an infinity of extensions of our starting theory T, as was illustrated previously for the case $T = \mathrm{PA}$. Each of the theories PA_i in the sequence PA, PA_1, PA_2,... is obtained by adding as a new axiom that the preceding theory is consistent. Given that PA is sound, all of the theories in this sequence are also sound. But we can say more, for if we form the theory PA_ω whose axioms are those of PA together with *all* of the consistency statements obtained in this way, P_ω is also sound. And the procedure can be continued, for we can now extend PA_ω, which is still subject to the incompleteness theorem, to a stronger theory $\mathrm{PA}_{\omega+1}$ by adding "PA_ω is consistent" as a new axiom.

What happens when this process is continued? Whenever we manage to define a theory like PA_ω, which is demonstrably sound, given that PA is sound, we will also be able to extend it to a stronger theory that is still sound. But then the question arises just when we can prove that a particular theory obtained in this way is in fact sound. The subject quickly becomes technical.

In the sequence of theories presented, each theory was obtained from the preceding one by adding as an axiom that the preceding theory is consistent. Gödel's remarks apply equally to theories obtained by adding to a theory T as a new axiom a *stronger* statement than "T is consistent," which still follows from the arithmetical soundness of T. For example, if every arithmetical theorem of T is true, the same is true for the theory obtained by adding to T the axiom "T is Σ-sound," or in other words, "every Goldbach-like statement disprovable in T is false." As in the sequence of extensions by consistency statements, we get a sequence of sound extensions of PA using this way of extending a theory in the sequence. The theories in this sequence will be logically stronger than the corresponding theories in the sequence obtained by adding consistency statements. Similar but stronger extension principles, known as *reflection principles*, can also be formulated.

In this book, we will return to the subject of inexhaustibility in Section 6.3, in connection with Gödelian arguments in the philosophy of mind. The mathematics of repeatedly adding consistency statements is not then at issue, but instead the argument that our inability to specify any formal system that exhausts our mathematical knowledge indicates that there is something essentially nonmechanical about our mathematical thinking.

6

Gödel, Minds, and Computers

6.1 Gödel and the UTM

Discussions of appeals to the incompleteness theorem in the philosophy of mind ("Gödelian arguments") sometimes refer to machines (computers), sometimes to formal systems. The two notions are interchangeable in these discussions, since for any formal system there is a way to program a computer to systematically generate the theorems of the system, and conversely, for any way of programming a computer to generate sentences in some formal language, there is a corresponding formal system which has those sentences among its theorems.

One of the most widespread misconceptions about the first incompleteness theorem is that Gödel's proof of it, when applied to a consistent system, shows the Gödel sentence of the system, unprovable in the system itself, to be true. Rudy Rucker, in *Infinity and the Mind* [Rucker 95, p.174], tells the following misleading story:

> The proof of Gödel's Incompleteness Theorem is so simple, and so sneaky, that it is almost embarrassing to relate. His basic procedure is as follows:
>
> 1. Someone introduces Gödel to a UTM, a machine that is supposed to be a Universal Truth Machine, capable of correctly answering any question at all.

2. Gödel asks for the program and the circuit design of the UTM. The program may be complicated, but it can only be finitely long. Call the program P(UTM) for Program of the Universal Truth Machine.

3. Smiling a little, Gödel writes out the following sentence: "The machine constructed on the basis of the program P(UTM) will never say that this sentence is true." Call this sentence G for Gödel. Note that G is equivalent to "UTM will never say G is true."

4. Now Gödel laughs his high laugh and asks UTM whether G is true or not.

5. If UTM says G is true, then "UTM will never say G is true" is false. If "UTM will never say G is true" is false, then G is false (since $G =$ "UTM will never say G is true"). So if UTM says G is true, then G is in fact false, and UTM has made a false statement. So UTM will never say that G is true, since UTM makes only true statements.

6. We have established that UTM will never say G is true. So "UTM will never say G is true" is in fact a true statement. So G is true (since $G =$ "UTM will never say G is true").

7. "I know a truth that UTM can never utter," Gödel says. "I know that G is true. UTM is not truly universal."

So far Rucker. We can invent a continuation of the story:

8. Gödel's jaw drops as UTM gravely intones, "I hereby declare that G is true." "But," Gödel manages to squawk, "you're supposed to always tell the truth." "Well," says UTM, "it seems I don't."

Suppose UTM is in fact a machine that always tells the truth, so that the final exchange can never take place. Gödel still hasn't demonstrated that he knows any truth that UTM can never utter. What he knows is only the implication "if UTM always tells the truth, then G is true." But this implication can be uttered by UTM as well. It is only if Gödel has somehow acquired the knowledge that UTM always tells the truth that he knows the truth of G. Being told that UTM is "supposed to be" a universal truth machine does not amount to knowing that UTM always tells the truth.

The argument put forward in [Lucas 61] to show that for any consistent formal system S, there is a true statement that we can prove but S can not, is invalid for the same reason. Lucas wrongly claims that "Gödel's theorem states that in any consistent system which is strong enough to produce simple arithmetic there are formulas which cannot be proved in the system, but which we can see to be true." As emphasized in earlier chapters, the theorem neither states nor implies any such thing. What we know on the basis of Gödel's proof of the incompleteness theorem is not that the Gödel sentence G for a theory S is true, but only the implication "if S is consistent then G is true." This implication is provable in S itself, so there is nothing in Gödel's proof to show that we know more than can be proved in S, as far as arithmetic is concerned. (A persistent reader of Rucker's book will learn as much from "Excursion Two" at the end of the book, which gives a presentation of the proof of the incompleteness theorem.) We do of course know the Gödel sentence of, for example, PA to be true, since we know PA to be consistent. Whenever we know a theory S to be consistent, we also know the truth of a statement not provable in S. But in those cases when we have no idea whether or not S is consistent, we also have no idea whether or not a Gödel sentence G for S is true, and if we merely believe or guess S to be consistent, we merely believe or guess G to be true.

Given that we have no basis for claiming that we ("the human mind") can outprove any given consistent formal system, a weaker claim could be made that there can be no formal system that *exactly* represents the human mind as far as its ability to prove arithmetical theorems is concerned. To test this claim, suppose we specify a theory T, say as ZFC with an added axiom of infinity (as described in Section 8.3). Can we use Gödel's theorem to disprove the claim that the arithmetical theorems of T are precisely those that the human mind is capable of proving? An attempt to do so fails for the same reason that the original Lucas argument fails. First, nothing in the incompleteness theorem rules out that every arithmetical statement provable in T can also be proved by the human mind. We may not at the moment know how to prove a given arithmetical theorem of T, but we can't use the incompleteness theorem to rule out the possibility that a proof acceptable to the human mind exists. Second, the incompleteness theorem does not rule out the possibility that every arithmetical statement provable by the human mind is provable in T. If we make a suitable choice of axiom of infinity, it will not be in any way evident to the human mind that T is consistent, and nothing in Gödel's theorem gives us any example of an arithmetical statement that we can prove or "see to be true" but which is unprovable in T.

To this standard argument, Lucas responds ([Lucas 96, p. 117]):

> The mind does not go round uttering theorems in the hope of tripping up any machines that may be around. Rather, there is a claim being seriously maintained by the mechanist that the mind can be represented by some machine. Before wasting time on the mechanist's claim, it is reasonable to ask him some questions about his machine to see whether his seriously maintained claim has serious backing. It is reasonable to ask him not only what the specification of the machine is, but whether it is consistent. Unless it is consistent, the claim will not get off the ground. If it is warranted to be consistent, then that gives the mind the premise it needs. The consistency of the machine is established not by the mathematical ability of the mind but on the word of the mechanist. The mechanist has claimed that his machine is consistent. If so, it cannot prove its Gödelian sentence, which the mind can none the less see to be true: if not, it is out of court anyhow.

These comments are somewhat odd, since they seem to set aside the question whether Gödel's theorem can be used to *disprove* the statement that T exactly represents human arithmetical ability, quite apart from whether anybody claims this to be the case. But suppose we adopt the attitude expressed by Lucas, that the point is what we can argue when faced with a "seriously maintained claim" that T represents the human mind. It is then noteworthy that Lucas only speaks of the machine being "warranted to be consistent." This is just like the "supposed to always tell the truth" in the story of the UTM. An earnest claim by "the mechanist" that the machine is consistent will not give Lucas any grounds for claiming the Gödel sentence of the machine to be in any sense humanly provable. Being assured that the machine is consistent gives no support to the claim that "the mind can see" that it is consistent. We need to distinguish between "We know that if T is consistent then G is true," which is true, and "If T is consistent, then we know that G is true," which we have no grounds for claiming. Of course, if Lucas has great confidence in the constructor of the machine, he will perhaps accept his claim and believe the machine to be consistent, and so also believe the Gödel sentence of the machine to be true, but this in no way amounts to proving the Gödel sentence or "seeing" it to be true.

6.2 Penrose's "Second Argument"

Roger Penrose, in his two books *The Emperor's New Mind* and *Shadows of the Mind*, has argued at length that Gödel's theorem has implications for a "science of consciousness." In *Shadows*, he presents a Gödelian argument ("Penrose's new argument," "Penrose's second argument") aiming to establish what the Lucas argument does not, that no machine can exactly represent the ability of the human mind to prove arithmetical theorems. The presentation of this argument in *Shadows* is fairly long and involved, but fortunately Penrose has set out the argument in its essentials in the electronic journal *Psyche* ([Penrose 96]):

> We try to suppose that the totality of methods of (unassailable) mathematical reasoning that are in principle humanly accessible can be encapsulated in some (not necessarily computational) sound formal system F. A human mathematician, if presented with F, could argue as follows (bearing in mind that the phrase "I am F" is merely a shorthand for "F encapsulates all the humanly accessible methods of mathematical proof"):
>
> (A) "Though I don't know that I necessarily am F, I conclude that if I were, then the system F would have to be sound and, more to the point, F' would have to be sound, where F' is F supplemented by the further assertion "I am F." I perceive that it follows from the assumption that I am F that the Gödel statement $G(F')$ would have to be true and, furthermore, that it would not be a consequence of F'. But I have just perceived that "if I happened to be F, then $G(F')$ would have to be true," and perceptions of this nature would be precisely what F' is supposed to achieve. Since I am therefore capable of perceiving something beyond the powers of F', I deduce that, I cannot be F after all. Moreover, this applies to any other (Gödelizable) system, in place of F."

Let us simplify further. Given a formal system F, Penrose invokes a corresponding mathematical statement (or, with suitable adjustments in the argument, a rule of inference) $IAMF$ such that

1. If $IAMF$ then $F + IAMF$ is consistent,

and furthermore,

2. I can prove (or "perceive"): if $IAMF$ then $F + IAMF$ is consistent.

In this formulation, the Gödel sentence for $F + IAMF$ has been replaced by the corresponding consistency statement, which is usually a good idea in these arguments, since the Gödel sentence for a theory S is equivalent in S to "S is consistent," which is less likely to promote confusion than the self-referential Gödel sentence. Penrose establishes (1) by way of the soundness of $F + IAMF$, but it is only essential to the argument that we can somehow prove (1). The next premise in the argument is

3. If $IAMF$ then for every A, if I can prove (or "perceive") A then F proves A.

It follows from (2) and (3) that if $IAMF$ then F proves "if $IAMF$ then $F + IAMF$ is consistent." But then $F + IAMF$ proves that $F + IAMF$ is consistent, so $F+ IAMF$ is inconsistent. This is incompatible with $IAMF$, by (1). So if $IAMF$ is true it is false, so $IAMF$ is false.

The argument is correct in the sense that not-$IAMF$ follows from (1)–(3) for any F to which the incompleteness theorem applies. Furthermore, it is easy to find mathematical statements $IAMF$ for which (1)–(3) hold: just take any statement that we can prove to be false. But of course the argument is pointless unless we can find such an $IAMF$ for which we have grounds for claiming

4. If I am F, in the sense that F encapsulates all the humanly accessible methods of mathematical proof, then $IAMF$.

Can an $IAMF$ for which (1)–(4) hold be found? Penrose comments:

> Of course, one might worry about how an assertion like "I am F" might be made use of in a logical formal system. In effect, this is discussed with some care in *Shadows*, Sections 3.16 and 3.24, in relation to the Sections leading up to 3.16, although the mode of presentation there is somewhat different from that given above, and less succinct.

This comment is wildly optimistic. It would lead too far to go into Penrose's presentation of the "second argument" in *Shadows*. Commentaries on Penrose's second argument in the literature (such as the contributions to the debate in Psyche, [Lindström 01], and [Shapiro 03]) differ considerably both in their attempted reconstructions of the argument and in their diagnoses of where the error lies. Any further contribution to these commentaries that might be included in this book would essentially only consist in a prolonged complaint, which few readers would be likely to find illuminating, that Penrose fails to substantiate the idea that there is any *IAMF* for which (1)–(4) hold.

Fortunately, there is little reason to enter into any sustained examination of Penrose's arguments on this point, for he comments himself ([Penrose 96, Section 4.2]) that

> ...I do not regard [the "second argument"] as the 'real' Gödelian reason for disbelieving that computationalism could ever provide an explanation for the mind—or even for the behavior of a conscious brain.

Instead, the argument that Penrose considers most persuasive, and to which he devotes Chapter 2 of *Shadows*, turns on our inability to fully specify our arithmetical knowledge.

6.3 Inexhaustibility Revisited

As Gödel emphasized, if we accept a formal system S as a correct formalization of part of our mathematical knowledge, we will also in the same sense, and with the same justification, accept an extension of that system obtained by adding as a new axiom "S is consistent." Since the resulting system is logically stronger than S, we conclude that we cannot specify any formal system S that exhausts our mathematical knowledge.

Penrose formulates an equivalent observation in somewhat eccentric terms as "Conclusion G" in [Penrose 94, p. 76]: "Human mathematicians are not using a knowably sound algorithm in order to ascertain mathematical truth." Instead of invoking the second incompleteness theorem, he applies Turing's proof of the unsolvability of the halting problem (see Section 3.3). Penrose's argument, slightly modified, is the following. If we *know* that every theorem of a formal system T of the form "P_i does not terminate for input i" is true, we can specify a statement of this form which we know to be true but is not a theorem of T. For we can find an e such that P_e is a program that given n looks for a theorem in T of the form

"P_n does not terminate for input n," and for this e, we know the statement "P_e does not terminate for input e" to be true but unprovable in T. In this sense there is no "knowably sound" theory T that proves every "humanly provable" statement of the form "P_n does not terminate for input n."

Unlike the Lucas argument, the observation that no "knowably sound" algorithm or theory exhausts our arithmetical knowledge is not based on any misunderstanding of the incompleteness theorem or its proof, and unlike Penrose's "second argument," it is not rendered opaque by various obscurities and uncertainties. But does it tell us anything about the human mind?

Penrose grants that his Conclusion G is compatible with the assumption that the human mind is in fact exactly equivalent to some formal system T as far as its ability to prove arithmetical statements is concerned, although we could in such a case not prove or perceive the consistency of T (and perhaps not formulate the axioms of T). He presents various arguments against this idea, and indeed the assumption that there is such a T has little to recommend it. To come to grips with the feeling that Conclusion G shows that the human mind is in some sense nonalgorithmic, we need to tackle head on the question whether it shows that it is in fact impossible *for us* to program a computer with all of our ability to prove arithmetical theorems.

It may seem that the answer is obvious, since what conclusion G states is precisely that we cannot specify any formal system that exhausts our mathematical knowledge. But there is no reason why programming a computer with all of our ability to prove arithmetical theorems should consist in specifying such a formal system. To emulate human mathematicians, the computer also needs to be able to apply precisely the kind of reasoning that leads us from accepting a formal system T as mathematically correct to accepting a stronger system as in the same sense correct.

Penrose, in considering this possibility ([Penrose 94, p. 81]), argues that whatever rules of reasoning of this kind with which we program the robot, the totality of statements provable by the robot will still be theorems of a formal system that we recognize as sound, and Conclusion G still applies: the robot has not been programmed with the sum total of our arithmetical knowledge and theorem-proving ability.

This argument is inconclusive because in fact the kind of reasoning that leads us from accepting a formal system T as mathematically correct to accepting a stronger system as mathematically correct covers a wide range of both formal and informal principles, some of which are evident and unproblematic, while others are less so. The case where we go from a theory T that we unhesitatingly accept as correct to the extension $T+$ "T is consistent" is simple and unproblematic. How could we incorporate this principle in trying to program a robot mathematician with all of our arithmetical knowledge? We start by programming the robot with some basic theory T of arithmetic. To include the extension principle, we might give the robot a button to press: whenever the button is pressed, the store of axioms available to the robot is extended by the statement that its earlier store of axioms is consistent. In searching for a proof of an arithmetical statement, the robot may press the button any number of times. Does this make the robot our mathematical equal? No, for since we recognize that all of the theories obtainable by repeatedly applying the above principle starting with T are correct, we also recognize that the union of these theories, which has all of the consistency statements obtained in this way as axioms, is correct, so we get a new consistency statement which can not be proved by the robot. But now we're applying a different principle: we go from a theory T that we unhesitatingly accept as correct to an extension of T by infinitely many consistency statements, which we also unhesitatingly accept as correct. We can program the robot to make use of this principle as well. But now we find that the robot still does not match our theorem-proving ability, for we can formulate stronger principles of extension.

As we continue to formulate stronger and more involved principles for extending a correct theory to a stronger theory that is still correct, we are confronted with a number of questions about what is or is not evident or mathematically acceptable, questions to which different mathematicians and philosophers will give different answers, and where many would say that there is no definite answer. To program a robot to perfectly emulate human mathematicians, we would need to give it a similar range of responses to these questions. If we manage to do this, the set of theorems provable by the robot using any of the formally defined extension principles will still indeed be computably enumerable, but we will have no grounds for the claim that we as human mathematicians can prove anything not provable by the robot. We will have succeeded in creating a robot that becomes just as confused and uncertain as humans do when pondering ever more complicated or far-reaching ways of extending a correct theory to a stronger correct theory.

The above brief description of what happens when we start thinking about various principles for extending a correct theory to a stronger correct theory can only be substantiated by going into technicalities. The interested reader can turn to [Franzén 04] for a sustained exposition.

6.4 Understanding One's Own Mind

A fairly common invocation of Gödel's theorem is illustrated by the following remark:

> According to Gödel's incompleteness theorem, understanding our own minds is impossible, yet we have persisted in seeking this knowledge through the ages!

Like so many similar references to the incompleteness theorem outside logic and mathematics, this comment, taken literally, is simply mistaken. "According to" means "as stated or implied by," and of course Gödel's incompleteness theorem neither states nor implies that understanding our own minds is impossible. But as in the case of other such statements, we need to understand the above reflection as inspired by the incompleteness theorem rather than as drawing any conclusion from it. Douglas Hofstadter's formulation of a similar reflection, in *Gödel, Escher, Bach* ([Hofstadter 79, p. 697]), has the virtue of making it explicit that the role of the incompleteness theorem is a matter of inspiration rather than implication:

> The other metaphorical analogue to Gödel's Theorem which I find provocative suggests that ultimately, we cannot understand our own minds/brains .... Just as we cannot see our faces with our own eyes, is it not inconceivable to expect that we cannot mirror our complete mental structures in the symbols which carry them out? All the limitative theorems of mathematics and the theory of computation suggest that once the ability to represent your own structure has reached a certain critical point, that is the kiss of death: it guarantees that you can never represent yourself totally.

Finding suggestions, metaphors, and analogies in other fields when studying the human mind is of course perfectly legitimate and may be quite useful. But it can only be a starting point, and actual theories and studies of the human mind would be needed to give substance to reflections like Hofstadter's. Metaphorical invocations of Gödel's theorem often suffer from the weakness of giving such satisfaction to the human mind that they tend to be mistaken for incisive and illuminating observations.

The question how to substantiate reflections like Hofstadter's, and whether it can be done, will not be considered in this book. However, this particular category of metaphorical applications of the incompleteness theorem merits some further comments.

In reflections such as those quoted, it is commonly the second incompleteness theorem that is explicitly or implicitly referred to. The inability of a formal system S to prove its own consistency is interpreted as an inability of S to sufficiently analyze and justify itself, or as a kind of blind spot. The system doesn't "understand itself."

To this, it may be objected that the metaphor *understates* the difficulty for a system to prove its own consistency. As commented on in Section 5.1, the unprovability of consistency is really the unassertibility of consistency. A system cannot truly *postulate* its own consistency, quite apart from questions of analysis and justification, although other systems can truly postulate the consistency of that system. An analogous difficulty for a human is that he cannot truly state that he never talks about himself, although other people can truly make this observation about him. The reason for this is not that the human mind cannot sufficiently analyze or justify itself, but that the very utterance of the statement "I never talk about myself" falsifies it.

Of course this analogy is not likely to strike anybody as provocative or suggestive, since it doesn't even have the appearance of saying anything about the human mind, but only makes a logical observation about the incompatibility of the content of a particular kind of assertion with the making of the assertion. So let us make an effort to draw a "conclusion" about the human mind by way of a comparison with formal systems like PA and ZFC. Instead of concluding that the human mind cannot understand itself, we will conclude that the human mind, if it is at all like these formal systems, is able to understand itself wonderfully.

For every finite subset of the axioms of PA, PA proves the consistency of that subset, and does so by shrewdly analyzing the logic of such finite parts of itself. Furthermore, PA proves

> For every finite subset of my axioms, I can prove that subset consistent

and also

> If every finite subset of my axioms is consistent, then I am consistent.

However, in order to avoid tedious paradoxes, PA cannot assert "If I can prove a finite subset of my axioms consistent, then that subset is consistent" (since in such a case PA would prove its own consistency, and therewith both the truth and the falsity of the Gödel sentence for PA). PA can of course also prove this fact about its own inability to consistently affirm such a principle. That is, PA proves

> If I can prove "For every finite subset M of my axioms, if I can prove M consistent then M is consistent" then I am inconsistent.

Inspired by this impressive ability of PA to understand itself, we conclude, in the spirit of metaphorical "applications" of the incompleteness theorem, that if the human mind has anything like the powers of profound self-analysis of PA or ZFC, we can expect to be able to understand ourselves perfectly.

"Going Outside the System"

It is often said that it is only by "going outside the system" that one can prove the Gödel sentence of a theory such as PA, an image that reinforces the idea that "a system cannot understand itself fully." This is correct in the sense that one cannot prove, or even truly postulate, the Gödel sentence of the system, or equivalently the consistency of the system, *in* the system, that is, by a proof formalizable in the theory. But the image of "going outside the system" is a bit too seductive, in that it suggests that there is some generally applicable way of viewing a system "from outside" so as to be able to prove things about it that are not provable in the system. We don't know of any such general method. Consider the theory T obtained by adding Goldbach's conjecture as a new axiom to PA. T is consistent if and only if Goldbach's conjecture is true, but we have no inkling of any method of "going outside the system" by which T might be proved or "seen" to be consistent.

7

Gödel's Completeness Theorem

7.1 The Theorem

A common source of confusion in connection with Gödel's incompleteness theorem is the fact that Gödel also proved, in his doctoral dissertation, an important result known as the

Completeness theorem for first-order logic (Gödel). *First-order logic (also known as predicate logic, first-order predicate logic, or first-order predicate calculus) is complete.*

As noted in Section 2.3, it is often said that predicate logic escapes the conclusion of the incompleteness theorem because it does not incorporate the "certain amount of arithmetic." For example:

> Gödel proposed that every formal system embodying a language complex enough that elementary number theory can be represented in terms of it is either incomplete or inconsistent. Most of the formal systems that we use in practice are both complete and consistent: e.g., first-order predicate calculus, Euclidean geometry. They aren't complex enough to qualify as "Gödelian."

Such comments are based on a misunderstanding due to an unfortunate overloading in logic of the term "complete." Euclidean geometry (suitably restricted and axiomatized) is indeed a complete theory in the sense of the incompleteness theorem, that is, every statement in the language of the theory is either provable or disprovable in the theory. This result, like the completeness of the elementary theory of the real numbers mentioned in Section 2.3, to which it is closely related, was proved by Alfred Tarski in the early 1930s. But first-order predicate calculus is not complete in this sense.

So what does the completeness theorem mean? Here we need to recall that a formal axiomatic system has a formal language, a set of axioms formulated in that language, and a set of rules of reasoning for drawing conclusions from those axioms. In the case of what is called a *first-order theory*, the language of the system is of a kind known as a *first-order predicate language*, and the rules of reasoning include (and often consist solely of) some version of the rules of first-order predicate logic. That first-order predicate logic is complete means that these rules suffice to derive from the axioms every sentence that is a *logical consequence* of the axioms. Since, conversely, every sentence derivable from a set of axioms using the rules of reasoning is a logical consequence of those axioms, it follows that the theorems of a first-order theory are precisely the sentences that are logical consequences of the axioms of the theory.

The present chapter contains a fairly brief explanation of first-order logic and a discussion of how the completeness theorem for predicate logic relates to the incompleteness theorem. A reader who finds the topic unrewarding need only retain from it that what is "complete," in the sense of the completeness theorem for predicate logic first proved by Gödel, is not an axiomatic theory, but the logical apparatus common to PA and ZFC and all formal systems known as first-order theories. That this logical apparatus is complete means that it suffices to deduce every logical consequence of the axioms of any such formal system.

First-order predicate logic has been an important part of logic since it was first formulated as a set of formal rules of inference in the latter part of the nineteenth century, chiefly by the German mathematician, logician, and philosopher Gottlob Frege. In Frege's formulation, the rules of first-order logic were not separated from the other parts of his logical system, and it was only in the early decades of the twentieth century that the distinctive character of first-order logic came to be understood. There are different formulations of the logical apparatus of first-order logic. Some versions include special logical *axioms*, so that one distinguishes between the "logical" and the "non-logical" axioms of a theory, where the logical axioms are common to all theories. Other versions have no logical axioms, but only logical rules of reasoning. In this book, for clarity and consistency, the logical apparatus of a theory has been assumed to be formulated wholly in terms of rules of reasoning.

First-Order Logical Consequence

To illustrate the notion of logical consequence, let us introduce a simple first-order predicate language. In such a language, we can formulate statements about the *individuals* in some *domain* of individuals, using a collection of *predicates* to express properties of and relations between individuals. In our example of a first-order language we can say that an individual has the property of being a *fnoffle*, and that one individual *glorfs* another. We say that "fnoffle" is a *unary predicate* and "glorfs" is a *binary predicate.* To form complex sentences we have at hand the logical constructions "if-then," "not," "or," "and," "if and only if," and we also use *variables* ranging over the domain of individuals to express statements of the form "for every (individual) x it holds that..." and "there is at least one (individual) x such that...."

Let us consider a few examples of sentence in this language, and how they might be formulated in ordinary language.

- There is at least one fnoffle: there is an x such that x is a fnoffle.
- No fnoffle glorfs itself: for every x, if x is a fnoffle it is not the case that x glorfs x.
- Every fnoffle is a fnoffle: for every x, if x is a fnoffle then x is a fnoffle.
- Every fnoffle is glorfed by at least one fnoffle: for every x, if x is a fnoffle then there is a y such that y is a fnoffle and y glorfs x.
- There is a fnoffle which glorfs every fnoffle: there is an x such that x is a fnoffle and for every y, if y is a fnoffle then x glorfs y.

In specifying the language, nothing is said about just what domain of individuals we are talking about, or what "x is a fnoffle" or "x glorfs y" is supposed to mean. The specification of the language is just a matter of syntax, and does not involve any questions of meaning. (See Section 2.4.) We can, however, make the following observation: no matter what domain of individuals we are talking about, no matter what subset of that domain is singled out by "is a fnoffle," and no matter what relation between individuals is specified by "glorfs," *if* the fifth sentence above is true, then the fourth sentence is also true. This is what is meant by saying that the fourth sentence is a *logical consequence* (in first-order logic) of the fifth sentence. We can also verify that the fifth sentence is *not* in this sense a logical consequence of the fourth sentence. For suppose we take the domain of individuals to contain only 0 and 1, interpret "is a fnoffle" as "is a member of the domain," and interpret "x glorfs y" as "x is identical with y." Then the fourth sentence is true, but the fifth sentence is false.

The third sentence above is an example of a *logically true* statement: it is true no matter what domain of individuals we choose, and no matter what "is a fnoffle" is taken to mean in that domain. In a formulation of predicate logic that includes logical axioms, those axioms are logically true statements.

Thus, informally speaking, to say that a sentence A in a first-order language is a logical consequence of a set M of sentences in that language means that for any domain of individuals, and for any specification of what the predicates used in the language mean when applied to individuals in that domain, if every sentence in M is true (when understood in accordance with this specification), then so is A. An *interpretation* of a first-order language is a structure consisting of a domain of individuals together with subsets of that domain and relations between elements in the domain corresponding to the predicates of the language. An interpretation is called a *model* of a first-order theory T if all the axioms of T are true when read using that interpretation. So another way of formulating the concept of logical consequence is that A is a logical consequence of the axioms of T if A is true in every model of T.

Soundness and Completeness of the Rules of Logic

The formal rules of logical reasoning used in a first-order theory have the property of being *sound* with respect to the notion of logical consequence. What this means is that anything that can be proved from a set of axioms using these rules of reasoning is also a logical consequence of the axioms in the sense defined. The *soundness theorem* for first-order logic establishes that this is the case. What the completeness theorem shows is that the converse holds: if A is in fact a logical consequence of a set of axioms, then there is a proof of A using those axioms and the logical rules of reasoning.

Recalling that a model of a theory is an interpretation in which all of the axioms of the theory are true, we can formulate the completeness theorem combined with the soundness theorem as the statement that for a first-order theory T,

> A sentence A is true in every model of T if and only if A is a theorem of T,

and also as the statement that for a first-order theory T,

> T has a model if and only if T is consistent.

No proof of the completeness theorem will be given in this book, but we can see that each of these two formulations of the theorems follows from the other. The second formulation follows from the first by taking A to be a logical contradiction "B and not-B," while the first formulation follows if we apply the second formulation to the theory $T+$ not-A. (The completeness theorem in fact holds also for the more liberal notion of "theory" explained in Section 4.3.)

We will take a closer look at these concepts in the particular case of PA.

7.2 PA as a First-Order Theory

The language of PA contains *function symbols*, which did not figure in the "fnoffle" example. Function symbols denote operations on individuals in the domain, and in the language of PA we have symbols $+$ for addition and $\times$ for multiplication. There is also a *constant*, the symbol 0 denoting the number zero, and a special symbol s for the *successor function*, which applied to a number n gives $n+1$. In the following we will use the notation $\underline{n}$, where n is any natural number, for the expression built up from 0 using n occurrences of s, so that $\underline{0}$ is 0, $\underline{1}$ is $s(0)$, $\underline{2}$ is $s(s(0))$, and so on. The expressions $\underline{n}$ correspond to the numerals used in ordinary mathematical language to denote natural numbers. There is only one predicate in the language of PA, namely $=$ denoting the equality relation. Those rules for $=$ that are not peculiar to arithmetic, for example the rule by which we conclude that $a = b$ if $b = c$ and $c = a$, are sometimes, but not always, regarded as part of the logical apparatus of the theory. Here we will follow this approach (taking PA to be formalized in "predicate logic with equality") and describe the specific content of PA in terms of its arithmetical axioms.

In the above presentation, the language of PA was specified together with the interpretation of the language that we normally use, known as "the intended interpretation" or "the standard model." But of course there are other interpretations of the language of PA, as will be commented on and illustrated below.

The axioms of PA fall into four groups. First, we have the axioms for the successor function, which state that 0 is not the successor of any number, and different numbers have different successors. Formulated in predicate logic, these are

For every x, it is not the case that $s(x) = 0$.

For every x, for every y, if $s(x) = s(y)$ then $x = y$.

The second group of axioms give the basic properties of the addition operation:

For every x, $x + 0 = x$.

For every x, for every y, $x + s(y) = s(x + y)$.

The axioms in the third group do the same for multiplication:

For every x, $x \times 0 = 0$.

For every x, for every y, $x \times s(y) = x \times y + x$.

The axioms for addition and multiplication are all we need to be able to prove every true statement of the form $\underline{n} + \underline{m} = \underline{k}$ or $\underline{n} \times \underline{m} = \underline{k}$. Using the axioms for the successor function we can also *disprove* every false statement of this form.

The final group of axioms consists of the *induction axioms*. For every property P of natural numbers expressible in the language of PA, there is an axiom stating that if 0 has the property P and $s(x)$ has the property P whenever x has the property, then for every x, x has the property P. Here, P may be defined using parameters, so that, for example, there is an axiom stating

For every y, if $0 + y = y + 0$ and it holds for every x that if $x + y = y + x$ then $s(x) + y = y + s(x)$, then for every x, $x + y = y + x$.

This particular induction axiom enters into the proof in PA that $x+y = y + x$ for every x and y.

Presburger Arithmetic

If we drop the multiplication symbol from the language of PA and delete the multiplication axioms, we get a theory known as *Presburger arithmetic* (after the Polish mathematician who introduced this theory in 1929). Presburger arithmetic is another example (in addition to the elementary theory of the real numbers mentioned in Section 2.3) of a theory which is *complete* without being at all trivial. Unlike PA, Presburger arithmetic can be finitely axiomatized, that is, it is possible to replace the infinitely many induction axioms with a finite number of other axioms and get a theory with the same theorems.

A Consistency Proof

In Section 2.8, in an argument against the idea of the incompleteness theorem leading to a "postmodern condition" in mathematics, it was observed that we can create an infinite tree of consistent variants of PA by omitting the axiom "for every x, $x + 0 = x$" and adding $0 + 0 = 0$ or its negation to PA, then to each of the two resulting theories add either $1 + 0 = 1$ or its negation, and so on.

Using the description given of PA and the soundness theorem for predicate logic, we can verify that all of these theories are indeed consistent, by showing that they all have models. In fact we can show that if we leave out the axiom "for every x, $x + 0 = x$," the result of an addition $n + 0$ can be anything. Suppose we replace the indicated axiom with an infinite number of axioms:

$$\underline{0} + 0 = \underline{n}_0, \underline{1} + 0 = \underline{n}_1, \underline{2} + 0 = \underline{n}_2, \ldots$$

where n_0, n_1, n_2,... is any computably enumerable sequence of numbers. We can define a model of the resulting theory as follows. The domain is again the set of natural numbers, the constant 0 is interpreted as denoting the number 0, and the successor operation has its ordinary interpretation. Note that this is enough to ensure that the induction axioms will still hold, whatever the interpretation of $+$ and $\times$. We change the interpretation of $+$ to denote the operation $+'$ defined by

$$k +' m = n_k + m.$$

It is still true that $k+'s(m) = s(k+'m)$. To make the axioms for $\times$ true, we now need to interpret $k \times m$ not as the ordinary product of k and m, but as $k \times' m$ where we *define* the operation $\times'$ on natural numbers by the equations

$$k \times' 0 = 0,$$
$$k \times' s(m) = k \times' m +' k.$$

Thus, all of the theories in the uninteresting infinite tree of theories described in Section 2.8 are consistent.

"True in the Standard Model"

The idea is sometimes expressed that instead of speaking of arithmetical statements as true or false, we should say that they are "true in the standard model" or "false in the standard model." The following comment illustrates:

> This is the source of popular observations of the sort: if Goldbach's conjecture is undecidable in PA, then it is true. This is actually accurate, if we are careful to add "in the standard model" at the end of the sentence.

The idea in such comments seems to be that if we say that an arithmetical statement A is "true" instead of carefully saying "true in the standard model," we are saying that A is true in every model of PA. This idea can only arise as a result of an over-exposure to logic. In any ordinary mathematical context, to say that Goldbach's conjecture is true is the same as saying that every even number greater than 2 is the sum of two primes. PA and models of PA are of concern only in very special contexts, and most mathematicians have no need to know anything at all about these things. It may of course have a point to say "true in the standard model" for emphasis in contexts where one is in fact talking about different models of PA.

7.3 Incompleteness and Nonstandard Models

An incorrect formulation of the first incompleteness theorem that is sometimes encountered is illustrated by the following:

> In any consistent theory T of a certain degree of complexity there will be a statement expressible in the language of T that is true in all models of T and yet not provable in T.

As we know from the completeness theorem, this statement is incorrect when we are talking about first-order theories like PA and ZFC. If a sentence A in the language of PA is true in every model of PA, it is provable in PA.

By the incompleteness theorem, the completeness theorem for first-order logic, and the consistency of PA, the theory obtained by adding to PA an arithmetization A of "PA is inconsistent" as a new axiom has a model. That is, there is a mathematical structure consisting of some set N' together with operations s', $+'$, and $\times'$ on the set N' such that all of the axioms of PA are true if we take s, $+$, and $\times$ to denote those operations on N', and furthermore A is also true. Since A is not true on its ordinary interpretation, that is, with the quantifiers taken to range over the natural numbers and $+$ and $\times$ denoting ordinary addition and multiplication, it follows that the model of PA given by N', s', $+'$, and $\times'$ is essentially different from the standard model.

We can in fact say a bit more. N' contains a part that is essentially the same as the ordinary natural numbers, namely the individuals in N' denoted by the terms 0, $s(0)$, $s(s(0))$,..., and for these individuals, $+$ and $\times$ have their usual meaning. But in addition, N' contains other members, which are usually called *infinite elements*. The reason for this terminology is that for any individual a in N' which is not the value of $\underline{m}$ for any natural number m, $\underline{n} < a$ is true in N' for every n. (Here we define $x < y$ to mean that $y = x + s(z)$ for some z.) This follows from the fact that it is provable in PA that for every x and y, either $x < y$ or $y < x$ or $x = y$, and that it is also provable, for every natural number n, that if $x < \underline{n}$ then $x = 0$ or $x = \underline{1}$ or ... or $x = \underline{n-1}$. So if a is not the value of any $\underline{m}$, we must have $\underline{n} < a$. A model of PA containing infinite elements is known as a *nonstandard* model. There are many different nonstandard models of PA, but essentially only one standard model. Although we can define the natural numbers in many different ways (for example, using set theory), as long as every natural number is the value of $\underline{n}$ for some n, the different models are *isomorphic*, meaning that they have exactly the same mathematical structure.

In a model of PA + "PA is inconsistent" there is an element e satisfying the condition expressed by the arithmetical formula that we introduce to formalize "x is the Gödel number of a proof in PA of a contradiction." Since e is an infinite element, it is not in fact a Gödel number. The operations $+'$ and $\times'$ in the nonstandard model are not multiplication and addition when applied to infinite elements, and the arithmetical definition of "x is the Gödel number of a proof in PA," when applied to an infinite element e, does not express that e is the Gödel number of anything.

Nonstandard models of arithmetic are studied in logic, but they also have a role to play in mathematics in general through the subject of *nonstandard analysis.* This is a subject created in the 1960s by the American logician Abraham Robinson, who used predicate logic to introduce a theory of the "infinitely small" within which the concepts and reasoning in calculus used by Newton and Leibniz, the creators of calculus, can be rigorously represented. Nonstandard analysis is a lively subject and a prominent example of how mathematical logic has been put to mathematical use. But it should not be assumed that we need to add a false existential axiom to PA in order to get a nonstandard model. On the contrary, the nonstandard models of arithmetic used in nonstandard analysis have exactly the same true arithmetical statements as the standard model. This can be achieved because the existence of nonstandard models of PA is not tied to incompleteness at all. Even true arithmetic (see Section 4.3) has nonstandard models. This follows from the completeness theorem. True arithmetic T, which is a theory only in a generalized sense, has every true arithmetical sentence as an axiom. Consider the theory obtained by adding to the axioms of T the new axioms $c > 0$, $c > \underline{1}$, $c > \underline{2}, \ldots$, where c is a new constant symbol. If this theory is inconsistent, T together with some finite number of the new axioms is inconsistent, since only finitely many axioms of T can be used in a proof. But T with any finite number of new axioms has a model, obtained by interpreting c as denoting a sufficiently large natural number, and hence is consistent. So T with all of the new axioms is also consistent and has a model by the completeness theorem, a model which contains an infinite element denoted by c. (In logic, this reasoning is known as an application of the *compactness theorem.*)

Thus, the difference between standard and nonstandard models of PA cannot be formulated in the language of elementary arithmetic. A nonstandard model may have exactly the same true sentences in the language of elementary arithmetic as the standard model. This fact has prompted many a commentator to agonize over how we distinguish between the standard and nonstandard models, but this is not a question directly connected with the incompleteness theorem, as noted, and so will not be taken up in this book.

8

Incompleteness, Complexity, and Infinity

8.1 Incompleteness and Complexity

In Section 2.2 it was commented that the common popular formulation of the incompleteness theorem as applying to any formal system of "sufficient complexity" is misleading at best, since there are very complex systems to which the theorem does not apply and very simple ones to which it does apply. Most often in such misleading formulations of the incompleteness theorem, "complexity" is perhaps used in an informal sense. In such a case it suffices to look at a system like Robinson arithmetic (defined in the Appendix) to see that very simple systems can encompass the "certain amount of arithmetic" needed for the incompleteness theorem to apply. In the other direction, it is a simple matter to formulate complete and consistent theories of impenetrable complexity. So if we use "complexity" in an informal sense, there is no correlation between complexity and incompleteness.

There is also a technical sense of "complexity" in logic, variously known as Kolmogorov complexity, Solomonoff complexity, Chaitin complexity (and various combinations of these names), algorithmic complexity, information-theoretic complexity, and program-size complexity. The most common designation is "Kolmogorov complexity"—as remarked in [Li and Vitanyi 97], this is probably a manifestation of the principle of "Them that's got shall get," since Kolmogorov is the most famous of these mathematicians. Again, the applicability of the incompleteness theorem does not turn on the complexity of a system in this technical sense, but there are connections between incompleteness and Kolmogorov complexity. In particular, the application of Kolmogorov complexity to prove the first incompleteness theorem has been widely popularized by Gregory Chaitin, one of the people who independently invented the concept.

Chaitin's Incompleteness Theorem

The concept of Kolmogorov complexity can be varied in many ways without affecting the aspects here to be considered. In extended theoretical treatments, the various technical concepts of complexity are usually applied to binary sequences, or bit strings, like 0110101010011, and the theory of Turing machines is used for technical definitions. Here, we will use instead the theory of computable properties of strings of symbols informally introduced in Chapter 3.

The basic idea is a simple one: we measure the complexity of a string s by the length, not of s itself, but of the shortest string containing information that enables us to *compute* the string s, in the form of a program that when executed outputs s, together with any input required by the program. For example, the complexity of the numerical string s consisting of the first one billion decimals of π is very small compared to the length of the string, since there are short programs for computing the decimals of π, and to get a program for producing the string s we need only add to such a program the instruction to stop after one billion decimals. We say that this string is highly *compressible*: its Kolmogorov complexity is small compared to its length. The opposite property is that of being highly *incompressible*, when the complexity of a string is close to its length. For any n greater than, say, 1000, the vast majority of strings of length n are highly incompressible. In particular for every n there is at least one string of length n which is maximally incompressible, having complexity n. (This follows by a simple counting argument: the number of programs of length smaller than n is at most $1 + 2 + 4 + \ldots + 2^{n-1} = 2^n - 1$, while there are 2^n strings of length n.) Incompressible strings are "typical," or "random," containing no pattern that makes it possible to give an algorithm for producing the string that is significantly shorter than the string itself. In the case of a maximally incompressible string, an algorithm for producing the string can do no better than explicitly listing the symbols in the string.

Thus, one would expect a genuinely random sequence of symbols (generated by observing physical processes like radioactive decay or the flow of lava lamps) to be incompressible. However, *proving* that a particular sequence is incompressible is a different matter.

Suppose T is a consistent formal system incorporating the usual "certain amount of arithmetic." Chaitin's incompleteness theorem states that there is then a number c depending on T such that T does not prove *any* statement of the form "the complexity of the string s is greater than c." Since there are true such statements, it follows that unless T proves false statements about complexity, there are statements of the form "the complexity of the string s is greater than c" that are undecidable in T.

We can formulate the essence of the argument showing this in semi-formal terms. Let $K(s)$ denote the Kolmogorov complexity of a string s. Suppose there are arbitrarily large n such that the consistent theory T proves $K(s) > n$ for some s, and that the string t gives sufficient information to enable us to computably generate the theorems of T. Choose n so that n minus the number of digits in a numerical string denoting n is several times larger than the length of t. To produce a string of complexity larger than n, we need only search through the theorems of T until we find the first theorem of the form $K(s) > n$. But this means that we can produce s using only the information t (giving T) together with the numerical string denoting n (to tell us what we are looking for), and the complexity of this combined information is much smaller than n, which yields a contradiction.

Before discussing this result further, a second and more formal proof of the theorem follows for readers who wish to understand the result in its technical aspects.

Recall from Section 3.3 the computable enumeration

$$P_0, P_1, P_2, \ldots$$

of the set of programs which expect one string as input, and when executed either deliver a string as output or else never terminate. For any string s, there are infinitely many i and w such that P_i with input w yields s as output. For example, if P_i is a program that simply returns its input as output, P_i with input s yields s as output. In defining Kolmogorov complexity (in this particular formulation), we are looking for the *shortest* way of specifying s by specifying a program and an input for that program which yields s as output. Thus we define $K(s)$ as the length of the *shortest* pair (i, w) such that P_i with input w yields output s.

It will be noted that the Kolmogorov complexity of a string thus defined depends on a particular choice of enumeration of programs. But it can be shown that for any two computable enumerations of programs that might be used, the resulting complexity measures $K_1(s)$ and $K_2(s)$ are essentially equivalent, in the sense that there is some constant c, depending on K_1 and K_2 but independent of s, such that $|K_1(s) - K_2(s)| < c$ for all s. This implies that the various results of a *qualitative* nature that we arrive at using the concept of Kolmogorov complexity, such as Chaitin's incompleteness theorem, are independent of exactly how we define complexity. (This also extends to various other ways of defining the concept, which in general yield measures of complexity that differ by a term logarithmic in the length of s.) We cannot, however, attach any great significance to *quantitative* results associated with a particular definition of $K(s)$.

The reference above to "the length of the pair (i, w)" calls for some words of explanation. Given any strings s and t, the pair (s, t) is defined as another string, from which s and t can both be read off. Thus given a pair, the two strings that compose the pair are uniquely determined. The exact definition of (s, t) doesn't matter here, but one way of defining the pair (s, t) is the following. Suppose s has length n. Then (s, t) consists of n occurrences of 1 followed by a 0, followed by s and then t. Thus, for example, if s is "abcdef" and t is "ghijk," (s, t) is the string

$$1111110\text{abcdefghijk}$$

The two strings in the pair (s, t) can be read off by first counting the number of 1s before the first 0, and then picking out s as the string consisting of the indicated number of symbols immediately following the 0, while t is the remainder of the string. This is called a *self-delimiting* representation of the pair formed by s and t. Note that if s has length n and t has length m, the length of (s, t) is $2n + m + 1$.

An important feature of Kolmogorov complexity is that a statement of the form $K(s) > n$ can be assumed to be a Goldbach-like statement. This is so because it can be formulated as the statement

> For every m and for every pair (i, w) of length smaller than or equal to n, it is not the case that m steps in the computation of P_i with input w yields s as output.

As a consequence, if a consistent theory proves $K(s) > n$, s does indeed have complexity greater than n.

To prove Chaitin's incompleteness theorem we also need a computable enumeration of the computably enumerable sets:

$$W_0, W_1, W_2, \ldots.$$

We define the set W_i as the set of strings s for which P_i terminates with s as input. W_i is computably enumerable, because the set of true statements of the form

$$P_i \text{ terminates after } n \text{ steps when given input } s$$

is computably decidable, so by generating all true statements of this form and delivering s as output for every true statement found, we get a computable enumeration of W_i. Conversely, every computably enumerable set E is identical with W_i for some (and in fact for infinitely many) i. This is so because if E is computably enumerable, E is the set of strings s for which the search for s in a computable enumeration of E terminates.

What if W_i is finite? The procedure described will in such a case not halt after every member of W_i has been generated, but will instead continue forever even after every element of W_i has already been delivered. This is in fact a necessary feature. There is no computable enumeration of the computably enumerable sets that allows us to determine which computably enumerable sets are finite, or only includes the infinite computably enumerable sets. This is why, in the definition of computably enumerable sets, we allow an enumeration of a finite set to continue forever.

For the proof of the theorem, we begin by defining a particular program P_e, which when given as input any pair (p, k), where p and k are numerical strings, starts looking for a sentence in W_p of the form "the complexity of w is greater than the length of the pair $(k, (p, k))$." If and when it finds such a sentence in W_p, the string w is output as the result of the computation.

Now, given a theory T, choose W_p as the set of theorems of T. Suppose P_e terminates for input (p, e) with result w. By the definition of complexity, the complexity of w is at most the length of $(e, (p, e))$. But if T is consistent, the complexity of w is greater than the length of $(e, (p, e))$. Thus, assuming T consistent, T does not prove any theorem of the form "the complexity of w is greater than c," where c is the length of the pair $(e, (p, e))$.

In the pair $(e, (p, e))$, e is independent of T, while p is an index for the set of theorems of T—to specify the theorems of T it suffices to specify p. It is reasonable to regard the minimal length of such a p as a measure of the complexity $K(T)$ of the theory T, and indeed it gives essentially the same measure of complexity as do other natural choices.

By suitably modifying and tweaking the definition of complexity, the incompleteness theorem can in fact be formulated as stating that, for some constant c independent of T, T does not prove any statement of the form $K(s) > n$ for any n greater than $K(T)+c$ (see [Chaitin 92]). The neatness of this formulation does not really add anything to the result, though, since neither the constant c nor the complexity measure used has any special significance. (See [van Lambalgen 89] and [Raatikainen 98].)

A consequence of Chaitin's incompleteness theorem is that the complement of the set of maximally incompressible strings, that is, the set of strings s with complexity smaller than the length of s, is a *simple* set in Post's sense (see Section 3.3). For the strings s with $K(s)$ smaller than the length of s can be computably enumerated by sifting through the statements of the form "m steps in the computation of P_i with input w yields output s." Furthermore, there cannot be any infinite computably enumerable set of maximally incompressible strings, since that would yield an infinite set W_p of true statements of the form "the complexity of s is greater than or equal to the length of s" and therewith a theory proving statements of the form "the complexity of s is greater than n" for arbitrarily large n. Thus, a consistent theory can prove only finitely many statements of the form "s is a maximally incompressible string."

When we inspect the proof that there are unprovable true statements of the form "the string s is maximally incompressible" in any consistent theory T, we see that the proof does not yield any explicit example of such a statement. If we had a general mechanism for producing, given a consistent theory, a true sentence of this form unprovable in the theory, we would be able to computably generate an infinite number of true such statements, which cannot be done. Thus, since "s is incompressible" is a Goldbach-like statement, we also cannot produce any statement of this form that we know to be *undecidable* in T. For any T, we can produce an n such that "s is an incompressible string of length n" is not provable in T for any s, but to know that a particular statement of this form is undecidable in T, we need to know that s is in fact an incompressible string of length n.

As noted by Chaitin (for example, in his paper "Gödel's Theorem and Information" [Chaitin 82]), the proof of his incompleteness theorem is related to what is known as *Berry's paradox*, in the way that Gödel's original proof is related to the paradox of the Liar. Berry's paradox (named after G. G. Berry, librarian at the Bodleian Library in Oxford and immortalized in a footnote in *Principia Mathematica*) consists in the observation that the smallest number not definable using fewer than a hundred words has just been defined using fewer than a hundred words. (Another proof of the incompleteness theorem exploiting Berry's paradox has been given by George Boolos [Boolos 89].)

Complexity as a Supposed Explanation of Incompleteness

Some of the comments that Chaitin has made about his incompleteness theorem in various places have given many of his readers the impression that the theorem sheds light on the grounds for incompleteness in general. Thus, in the abstract of [Chaitin 82], Chaitin states that

> Gödel's theorem may be demonstrated using arguments having an information-theoretic flavour. In such an approach it is possible to argue that if a theorem contains more information than a given set of axioms then it is impossible for the theorem to be derived from the axioms. In contrast with the traditional proof based on the paradox of the liar, this new viewpoint suggests that the incompleteness phenomenon discovered by Gödel is natural and widespread rather than pathological and unusual.

It is unclear what Chaitin intends by "it is possible to argue...," or in the body of the paper by the remark that "I would like to be able to say that if one has ten pounds of axioms and a twenty-pound theorem, then that theorem cannot be derived from those axioms." By "contains more information," Chaitin means "has greater Kolmogorov complexity," and apparently what Chaitin has in mind in these comments is his incompleteness theorem. This theorem, however, doesn't say anything about the complexity of the theorems of a theory, but instead deals with theorems that are statements *about* complexity. It is indeed the case that a true statement of the form "$K(s) > n$" must itself have complexity greater than n, since the string s can be extracted from it. But this does not mean that it is the *complexity* of the statement that accounts for its being unprovable in the theory. It is by no means the case that a theorem cannot have greater complexity than the axioms from which it is derived. For example, if we have as our only axiom "for every string x, $x = x$," this yields a theory with very low complexity. Among the theorems of this theory, however, are the statements of the form "$s = s$" for any specific string s, and thus the Kolmogorov complexity of the theorems is unbounded. So while it may be "possible to argue" that the theorems of a theory cannot be more complex than its axioms, and while it may be that Chaitin "would like to be able to say" that a fat theorem cannot be derived from skinny axioms, such statements have no apparent justification.

In fact Chaitin in [Chaitin 92] notes as much, and proposes an "improved version" of the "heuristic principle" that the complexity of a theorem cannot exceed that of the axioms:

> One cannot prove a theorem from a set of axioms that is of greater complexity than the axioms *and know* that one has done this. I.e., one *cannot realize* that a theorem is of substantially greater complexity than the axioms from which it has been deduced, if this should happen to be the case.

Chaitin suggests no justification for this "improved version," and it is not obvious what such a justification might look like. For any string s, we can prove $s = s$ from the axiom "for every string x, $x = x$." The quoted principle implies that we can never know that any string s has higher complexity than the string "for every string x, $x = x$," a baseless claim.

Chaitin concludes with the reflection that "Perhaps it is better to avoid all these problems and discussions by rephrasing our fundamental principle in the following totally unobjectionable form: a set of axioms of complexity N cannot yield a theorem that asserts that a specific object is of complexity substantially greater than N." This is unobjectionable as a formulation of his incompleteness theorem in the form "T does not prove $K(s) > n$ for any n greater than $K(T) + c$," if we keep in mind that it depends on a particular choice of complexity measure and that the constant c may be enormous, rendering the description "not substantially greater" moot. This is a special case of the incompleteness phenomenon, one that we can understand on the basis of informal arguments such as the one presented earlier.

As for the contrasting of the "new viewpoint" with "the old proof based on the paradox of the liar," proofs of the first incompleteness theorem based on computability theory have been around for a long time, as Chaitin himself notes. In particular, the MRDP theorem (see Section 3.3) shows that every theory to which the incompleteness theorem applies leaves undecided infinitely many statements of the form "the Diophantine equation $D(x_1,\ldots,x_n) = 0$ has no solution." It is indeed important to emphasize that undecidable arithmetical sentences need not be formalizations of odd self-referential statements, but this point is fully illustrated by the undecidability of statements about Diophantine equations.

8.2 Incompleteness and Randomness

Incompressible strings are also called (Kolmogorov) *random* strings. While randomness in this sense is a matter of degree, there is also a notion of random *infinite* sequences of symbols or numbers, on which a sequence is random or not random, without gradation. The definition of this property is rather more technical than the definition of the Kolmogorov complexity of strings and will not be given here. There is a particular such random infinite bit sequence (that is, a sequence of occurrences of 0 and 1) defined and introduced by Chaitin, called "the halting probability" or Ω (Omega), with several interesting properties. The sequence Ω is the *limit* of a certain computable enumeration $r_1, r_2, r_3, \ldots$ of finite strings of bits, in the sense that the nth bit of Ω is i if and only if there is some k such that the nth bit of r_m is i for every $m > k$. Ω itself is not computable: there is no effective enumeration of the bits of Ω. We can compute the strings r_1, r_2, $r_3, \ldots$, and we know that for any bit position n in Ω, the strings r_1, r_2, $r_3, \ldots$ will eventually stabilize so that every later string r_k has the same nth bit, which is also the nth bit of Ω, but we cannot in general decide at what point the nth bit has been arrived at.

Chaitin proves the stronger statement that only finitely many statements of the form "the nth bit of Ω is i" can be correctly decided in any one formal system, and he also proves, using a suitable definition of complexity, that a formal system of complexity n can determine at most $n + c$ bits of Ω, for some constant c.

So what is the significance of Ω for incompleteness? Chaitin, in the preface to his book *The Unknowable*, states

> In a nutshell, Gödel discovered incompleteness, Turing discovered uncomputability, and I discovered randomness—that's the amazing fact that some mathematical statements are true for no reason, they're true by accident.

The statements that Chaitin is referring to would appear to be the true statements of the form "The nth bit of Ω is i." "True for no reason" suggests that these statements are in some absolute sense unprovable by "reasoning," and indeed Chaitin elsewhere ("Irreducible Complexity in Pure Mathematics") amplifies:

> In essence, the only way to prove such mathematical facts is to directly assume them as new mathematical axioms, without using reasoning at all.

We will take a closer look at this idea below, but first we need to take notice of the fact that Chaitin also calls these statements "true by accident," emphasizing the "randomness" of Ω. In his recent book *MetaMath*, he comments

> Now we're really going to get irreducible mathematical facts, mathematical facts that "are true for no reason," and which simulate in pure math, as much as is possible, independent tosses of a fair coin: It's the bits of the base-two expansion of the halting probability Ω.

This suggests that statements of the form "The nth bit of Ω is i" are "true for no reason" in a sense analogous to that in which it may be "true for no reason" that, say, an atom decays at a particular time. When we say that radioactive decay is "truly random," we mean that there is no mechanism that causes an atom to decay at a particular moment and that a statistical description of radioactive decay tells us all there is to know about it. But of course Ω is not defined in terms of any physical experiments, and we have no conception of any "mathematical mechanism" or "laws of mathematics" by which some mathematical statements are caused to be true whereas others just happen to be true. If we are not talking about proofs and reasoning, but about what makes a statement true, all we can say is that Goldbach's conjecture, if true, is true because every even number greater than 2 is the sum of two primes, a statement "the nth bit of Ω is i," if true, is true because i is the limit of the nth bit of the mathematically defined sequence $r_1, r_2, r_3, \ldots$ of finite strings of bits, and so on. The mere fact that Ω is called a "random sequence" does not confer any automatic meaning on a claim that the bits of Ω are random in any sense analogous to the randomness that may exist in nature.

Indeed, Chaitin himself goes on to emphasize that Ω is a mathematically defined sequence, and that it may be preferable to use some other term than "random" to describe it. He suggests "irreducible":

> In other words, the bits of Ω are logically irreducible, they cannot be obtained from axioms simpler than they are. Finally! We've found a way to simulate independent tosses of a fair coin, we've found "atomic" mathematical facts, an infinite series of math facts that have no connection with each other and that are, so to speak, "true for no reason" (no reason simpler than they are).

This brings us back to the idea that these statements can only be postulated, not proved in any more interesting sense, which in the above passage seems to be based on their not being "obtainable from simpler axioms."

Let's take a closer look at this idea. Given any concept of "simpler," if the degree of simplicity of a string is given by a natural number, it is trivially the case that there are true arithmetical statements that cannot be logically derived from simpler true statements. For example, the simplest statement among the true statements that logically imply "$0 = 0$" cannot be logically derived from any simpler true statement. Indeed, on many natural measures of simplicity, "$0 = 0$" itself will have this property. The "amazing fact" Chaitin speaks of thus cannot be just the fact that there *exist* true statements that cannot be deduced from simpler true statements.

There are of course significant differences between "$0 = 0$" and statements of the form "The nth bit of Ω is i." In particular, such statements are not trivially or obviously true, but instead pose mathematical problems and call for a proof. But nor is there any apparent basis for claiming that they "cannot be obtained from axioms simpler than they are." The result that a formal system of complexity n can determine at most $n + c$ bits of Ω doesn't tell us that, for example, "The 1000th bit of Ω is 0," if true, can only be proved by assuming the statement as an axiom. Which particular statements about the bits of Ω that can be proved in a given system depends on the system and on the details in the definition of Ω. In particular, for any n, the parameters in the definition of Ω can be chosen so as to allow us to prove every true statement of the form "the kth bit of Ω is i" for $k < n$. For a natural choice of those parameters, Calude et al. [Calude et al. 01] computed the first 64 bits of Ω.

Let us consider instead statements of the form "the first n bits of Ω are given by s," where s is a bit string of length n. The randomness of Ω implies that there is some constant c such that the complexity of the string $\Omega_{1:n}$ of the first n bits of Ω, using a suitable definition of complexity, is always no smaller than $n - c$ for some c independent of n. Thus for large enough n, $\Omega_{1:n}$ is incompressible (to any degree we care to specify), and a true statement of the form "the first n bits of Ω are given by s" cannot be generated by an *algorithm* using as input a string (essentially) shorter than n. But we know that statements of enormous complexity can, in general, be *proved* from very simple axioms, even though there is no algorithm which yields any of those statements given a simple input. A special argument would be needed to show that sufficiently long statements of this particular form can only be proved using axioms of the same complexity, that is (since these statements are incompressible) of the same length. Even if we could show this to be the case, there is no apparent basis for the idea that if *reasoning* can establish the truth of a statement B using axioms A, then the axioms A must be shorter than the statement B (an idea that neatly complements Chaitin's other, contrary, "heuristic principle" that the complexity of a theorem cannot exceed that of the axioms from which it is derived). Chaitin does not present any argument for this idea, and it is not in agreement with our experience. Consider the statement

There is a prime p such that $10000^{10000} < p < 2 \times 10000^{10000}$.

Reason is by no means powerless to prove this statement—it follows from Erdös' nice elementary proof of Chebyshev's theorem that there is always a prime between n and $2n$. But for the proof we need to use various arithmetical axioms that have greater length and complexity, on any natural measure, than the theorem we wish to prove. A justification for the idea that statements of the form "the first n bits of Ω are given by s" (for large enough n) can only be proved by postulation, not by reasoning, must be sought in some other aspect of these statements than the incompressibility of $\Omega_{1:n}$.

Of course we know that reasoning formalizable in, for example, ZFC cannot prove any statement of this form for large enough n. (In fact, as Robert Solovay has shown, by suitably tweaking Chaitin's definition of Ω, "large enough n" can mean "any $n > 0$.") But to conclude that "reasoning" as such is powerless to establish these statements, we need some analysis or theory of reasoning in general (involving statements of arbitrary length), which is completely lacking in Chaitin's writings.

This is not to say that we have any grounds for claiming that statements about the bits of Ω, or statements of the form "$K(s) > n$," *are* always "true for a reason" in any theoretical sense. Again, we simply don't have any theoretical basis for reasoning about what can or cannot be proved or explained or given a reason in any absolute sense. Chaitin's claim to have discovered "the amazing fact that some mathematical statements are true for no reason" has no apparent content, but seems rather to be based on the general associations surrounding the word "random." The instances of arithmetical incompleteness found using the theory of Kolmogorov complexity are still of considerable interest.

8.3 Incompleteness and Infinity

Varieties of Incompleteness

Many incomplete formal systems are designed to be incomplete, since their purpose is to specify certain features of a particular class of mathematical structures, the properties of which otherwise vary. A typical example is the elementary theory of groups, which defines what is meant by "a group": it is a mathematical structure in which the axioms of the theory are true. The theory is incomplete, since, for example, some groups have a property called commutativity while others do not, and not all groups have the same number of members. Usually, theories introduced to characterize a class of mathematical structures do not satisfy the conditions for Gödel's theorem to apply, and their incompleteness is fairly obvious from their definition.

However, even if we restrict consideration to theories that are not used to define a certain class of mathematical structures, but are formalizations of parts of our mathematical knowledge, theories to which the incompleteness theorem applies, there are significant differences between the various incompleteness results proved in logic. Let us consider ZFC, Zermelo-Fraenkel set theory with the axiom of choice, which is a theory within which most of the mathematics done today is formalizable. ZFC is incomplete in several dimensions. (In the following comments, the arithmetical soundness of ZFC, or in other words the truth of all arithmetical theorems of ZFC, will be taken for granted.)

First, since the incompleteness theorem applies to ZFC, the arithmetical component of ZFC is incomplete. In particular, ZFC does not prove its own consistency. From the MRDP theorem, we know that there are undecidable statements in ZFC of the form "The Diophantine equation $D(x_1,\ldots,x_n) = 0$ has no solution," but the arithmetical statements *known* to be undecidable in ZFC (given the arithmetical soundness of ZFC) do not include any such problems that have been posed by mathematicians, or any problems about primes or other such matters of general mathematical interest. Although it cannot be excluded on general logical grounds that, for example, the twin prime conjecture is undecidable in ZFC, there is no reason whatever to believe this to be the case, and a proof that the twin prime conjecture is undecidable in ZFC would be a mathematical sensation comparable to the discovery of an advanced underground civilization on the planet Mars.

Second, the axioms of ZFC prove the existence of a profusion of infinite sets, but leave many statements about these infinite sets undecided. Most famous of these is Cantor's continuum hypothesis (CH), which states that every infinite subset of the set of real numbers either has the same cardinality as the set of real numbers itself or is countably infinite. (We don't need to go into what this means.) The continuum hypothesis is known to be undecidable in ZFC through work by Gödel and Paul Cohen. Cohen, building on Gödel's work in set theory, introduced a method called *forcing*, which has been extremely successful in proving statements about infinite sets to be unprovable in ZFC. This method has nothing to do with Gödel's incompleteness theorem. CH is typical of statements shown to be undecidable in ZFC using these methods, in that neither CH nor not-CH implies any *arithmetical* statement that is not already provable in ZFC. (Also, only the consistency of ZFC needs to be assumed in proving that CH is undecidable in ZFC.)

In his fundamental set-theoretical work establishing, among other things, the consistency of CH with the axioms of ZFC, Gödel introduced a set-theoretical principle known as "the axiom of constructibility," usually written "V = L." V = L settles all, or practically all, of the known undecidable statements in the second category (although we also know that it has no *arithmetical* consequences which are not already provable in ZFC). Furthermore, Gödel proved that V = L is consistent with the axioms of ZFC. (Thus, for undecidable statements in the second category, showing that they cannot be disproved in ZFC is often done by showing that they follow from V = L.) A reader who wishes to understand why V = L has nevertheless not been incorporated into the standard axioms of set theory can turn to Shelah's "Logical Dreams" [Shelah 03], which contains a more technical and searching discussion of several issues touched on in this final part of the book.

The third dimension of incompleteness of the axioms of ZFC is found in the fact that they do not decide just how "large" infinities exist. There is a family of statements known as "axioms of infinity" that have been intensively studied in set theory since the 1960s. These statements assert the existence of sets that, if they exist, must be "very large." In the case of axioms of infinity, the consistency of ZFC only implies that an axiom of infinity A is not provable in ZFC, that is, that ZFC + not-A is consistent. The consistency of the theory ZFC + A may be more or less problematic, even given the truth of the axioms of ZFC. There is a broad distinction between axioms of infinity that assert that the universe of sets has certain closure properties and therefore contains very large sets, and axioms of infinity that isolate some property of the smallest infinite set, the set of natural numbers, and simply state that there are larger infinite sets with a corresponding property. While the former category of axioms of infinity ("weak axioms of infinity") seem convincing to many, and justifiable on the basis of informal considerations, probably very few people would claim that it is at all evident that axioms of infinity in the second category ("strong axioms of infinity") are even consistent with the axioms of ZFC. Strong axioms of infinity are often inconsistent with V = L and are logically stronger than the axioms in the first category, although no axiom of infinity is known which settles the continuum hypothesis.

This third dimension of incompleteness of ZFC, and the study of axioms of infinity to which it has given rise, may seem a highly esoteric and specialized aspect of mathematics, and so it is. But at the same time, it is relevant to the mathematics of the natural numbers.

The Gödelian Connection

The reason for this is a remarkable connection between the incompleteness of ZFC in regard to the existence of very large infinite sets and the arithmetical incompleteness revealed by Gödel's incompleteness theorem: extensions by axioms of infinity always have arithmetical consequences not provable in the theory they extend, and the stronger the axiom of infinity, the more new arithmetical theorems it implies.

The connection can be understood on the basis of the incompleteness theorem. The weakest axiom of infinity is actually part of ZFC: it states that there exists an infinite set. Using this axiom, we can prove in ZFC the existence of a model of the remaining axioms of ZFC, and therewith prove the consistency of the theory $\text{ZFC}^{-\omega}$ obtained by leaving out the axiom of infinity from ZFC. $\text{ZFC}^{-\omega}$ is subject to the incompleteness theorem (and is equivalent, in its arithmetical part, to PA) and so does not prove its own consistency. Thus, we see that introducing the axiom "there is an infinite set" yields new arithmetical theorems, in particular the consistency of $\text{ZFC}^{-\omega}$, and therewith of PA. (This basic example is in a way misleading, since ZFC is in fact arithmetically very much stronger than $\text{ZFC}^{-\omega}$.)

Stronger axioms of infinity are not as easily formulated, but the basic connection between infinity and arithmetic follows the same Gödelian principle. Thus, the next axiom of infinity, asserting the existence of what is known as a strongly inaccessible cardinal, implies among other things that every arithmetical theorem of ZFC is true. An axiom that asserts the existence of an inaccessible cardinal is a "weak" axiom of infinity in the sense explained above. "Strong" axioms of infinity are far from transparent as regards their consequences, but investigations have shown that extensions of ZFC by the stronger axioms imply the arithmetical soundness of extensions of ZFC by the weaker axioms.

From a philosophical point of view, it is highly significant that extensions of set theory by axioms asserting the existence of very large infinite sets have logical consequences in the realm of arithmetic that are not provable in the theory that they extend. Again, however, no arithmetical problem of traditional mathematical interest is known to be among the new arithmetical theorems of extensions of ZFC by axioms of infinity, and it is highly desirable to find arithmetical statements of interest outside the special field of logic which are not decidable in ZFC but can be proved using axioms of infinity.

The Paris-Harrington Theorem

A first step in this direction was taken in 1977, when it was proved that a certain combinatorial principle is undecidable in PA. The great interest of this result lay in the fact that the combinatorial principle in question did not refer to Gödel numbers or formal theories, but was a seemingly insignificant strengthening of a well-known mathematical principle. It is worthwhile to take a closer look at this result, to obtain a broad understanding of the far-reaching extensions of this approach that will be described at the end of this chapter.

Ramsey's theorem is a fundamental result in finite mathematics, proved by Frank Plumpton Ramsey in 1928. To formulate this theorem we need some definitions. Suppose we have a finite set A of numbers (or indeed of anything else). An *n-element subset* of A is a set B with n elements, all of which are taken from A. An *m-partition* of the n-element subsets of A sorts those subsets into m categories $C_1, \ldots, C_m$ so that each n-element subset of A belongs to one and only one of these categories. Finally, a subset H of A is *homogeneous* for the given partition if there is some particular category C_i in the partition such that *every* n-element subset of A all of whose elements are taken from H falls in the category C_i.

Here is a concrete example. If A is the population of a city, we can partition the 5-people subsets of A into two categories: those groups of five people who get on well together, and those who do not. A homogeneous subset is a set H of people such that either any five people taken from H get on well together, or else any five people taken from H do not get on well together.

Ramsey's theorem is about the existence of homogeneous subsets. The theorem (or more specifically "the finite Ramsey theorem") states that for any n, m, and k there is a number p such that if A contains at least p elements, there is, for any m-partition of the n-element subsets of A, a homogeneous subset of A containing at least k elements. In the example (choosing $k = 1000$), if we specify a large enough population, we know that for any population at least that size, there is a set of 1000 people in the population such that either any five people from that subset get on well together, or else any five people from that subset do not get on well together.

The finite Ramsey theorem is a theorem about finite sets, but by representing finite sets of numbers as numbers, it can be formulated as an arithmetical statement. It is then provable in PA. The *Paris-Harrington theorem* is about a slight modification of the conclusion of the finite Ramsey theorem. We say that a set A of numbers is *relatively large* if the number of elements in A is greater than the smallest number in A. It can be proved that Ramsey's theorem holds also if we require the homogeneous set H to be relatively large. But, the Paris-Harrington theorem shows, this is *not* provable in PA.

While the Paris-Harrington theorem can be proved using different methods, the incompleteness in PA that it reveals is an instance of Gödelian incompleteness, in the sense that the strengthened Ramsey theorem is in fact equivalent in PA to "PA is Σ-sound." Thus, this is an instance of incompleteness in PA that can be remedied by proving the soundness of PA, and in particular can be proved on the basis of an axiom of infinity.

Later Developments

The Paris-Harrington theorem suggests that similar combinatorial principles in the mathematics of finite sets could be found which are unprovable in ZFC but can be proved using axioms of infinity. Harvey Friedman has produced a large body of work in this direction, showing various combinatorial principles to be equivalent to the consistency or the Σ-soundness of extensions of ZFC by various axioms of infinity. These results are similar to the Paris-Harrington theorem in that seemingly minor modifications of combinatorial principles provable in PA result in principles that imply the consistency of strong theories. The combinatorial principles formulated in Friedman's theorems are somewhat recondite, and it is an open question to what extent they can be put to use in "ordinary mathematics." Also, when these principles imply the consistency of "strong" axioms of infinity, it is not at all clear why the results should prompt us to accept the combinatorial principles. Gödel expressed the idea that the consequences of axioms of infinity in the realm of finite mathematics may be so rich and illuminating as to prompt us to accept axioms of infinity that are not evident considered as assertions about the existence of infinite sets. Friedman's work is clearly relevant to this idea, and it remains to be seen what its eventual mathematical outcome will be. An accessible discussion of issues connected with new axioms in mathematics can be found in [Feferman et al. 00].

A

Appendix

A.1 The Language of Elementary Arithmetic

The main purpose of this appendix is to give a formal definition of the concept of a Goldbach-like arithmetical statement and to comment on the significance of Rosser's strengthening of Gödel's original formulation of the first incompleteness theorem. A reader who would like to see a full proof of the incompleteness theorem can profitably turn to [Smullyan 92], which contains a wealth of information and does not presuppose any prior knowledge of logic.

The language of elementary arithmetic was defined in Section 7.2. To recapitulate, with some additional terminology, we begin with the *terms*. These are built up from *variables* x, y, z,..., the numeral 0, symbols $\times$ for multiplication and $+$ for addition, and finally the symbol s, denoting the successor function, which takes a number n to $n+1$. Thus the natural numbers 0, 1, 2, 3... are denoted in this language by the (formal) numerals 0, $s(0)$, $s((0))$, $s(s(s(0)))$,.... We write $\underline{n}$ for the numeral with value n. The variables are used the way one does in mathematics, and in particular to form terms like $x + s(0)$, $(x + y) \times z$, and so on, which have a natural number as value once values have been assigned to the variables in the term. For example, if x is assigned the value 8, y the value 0 and z the value 2, $(x + s(y) \times z)$ has the value 18. *Formulas* are formed in the language from *equalities* $s = t$, where s and t are terms, using logical operators (called *connectives*) "not," "or," "if-then," and "if and only if," together with the *universal quantifier* "for every natural number x" and the *existential quantifier* "for at least one natural number x."

Thus we can say in this language such things as

> For every natural number x, there is a natural number y such that $y = x + s(z)$ for some z, and $y = s(s(w))$ for some w, and for all natural numbers u and v, if $y = u \times v$ then $u = s(0)$ and $v = y$ or $u = y$ and $v = s(0)$,

which, as we shall see, is a way of expressing that there are infinitely many primes. We are free to extend the language of elementary arithmetic by introducing *defined* operations and relations, where the definitions are given using language already introduced. Thus, writing $x < y$ instead of "$y = x + s(z)$ for some z," and "y is a prime" instead of "$y = s(s(w))$ for some w, and for all natural numbers u and v, if $y = u \times v$ then $u = s(0)$ and $v = y$ or $u = y$ and $v = s(0)$," the above statement becomes the more readily intelligible

> For every natural number x, there is a natural number y such that $x < y$ and y is a prime.

In talking about formulas, a recurring distinction is that between *free* and *bound* occurrences of variables. Bound occurrences of variables are those that are governed by a quantifier. Thus, in "$y = s(s(w))$ for some w, and for all natural numbers u and v, if $y = u \times v$ then $u = s(0)$ and $v = y$ or $u = y$ and $v = s(0)$," all occurrences of the variables w, u and v are bound. These variables are only used in the formula as a means of expressing the logical constructs "for every number" and "for some number." The occurrence of the variable y in the indicated formula is free, which means that the formula as a whole expresses a condition on y, in this case that y is a prime.

A formula A with one free variable x is often written $A(x)$, and $A(t)$ then stands for the formula obtained by replacing x with t. A *sentence* in the language is a formula containing no free variables. Thus a sentence expresses a statement, which it makes sense to speak of as true or false, while a formula with free variables x, y,... expresses a condition on x, y,..., which some numbers may satisfy while others do not. If $prime(y)$ is the indicated formula expressing that y is a prime, $prime(\underline{17})$ is a true sentence expressing that 17 is a prime, and $prime(\underline{15})$ is the false sentence expressing that 15 is a prime.

It is a remarkable fact, discovered by Gödel, that using only addition and multiplication, we can *define* exponentiation and all of the other usual operations on numbers in this language. That exponentiation is definable means that there is a formula $exp(x, y, z)$ such that for all natural numbers m, n, k, the sentence $exp(\underline{m}, \underline{n}, \underline{k})$ obtained by substituting the numerals denoting these numbers for the variables x, y, z is true if and only if $k = m^n$. Furthermore, using this definition, PA proves the basic rules of exponentiation, that is $x^0 = 1$ and $x^{y+1} = x^y \times x$ for all numbers x, y. (From this together with the induction axioms other rules for exponentiation follow.) Thus, in discussing the subject of elementary arithmetic in a philosophical or logical context, we can restrict ourselves to addition and multiplication and the axioms of PA.

A Bit of Symbolism

Introducing some standard logical symbolism, we write $\forall x$ for "for every x," $\exists x$ for "there is an x such that," $\neg A$ for "it is not the case that A," $A \vee B$ for "A or B," $A \wedge B$ for "A and B," $A \supset B$ for "if A then B," and $A \equiv B$ for "A if and only if B."

Using these symbols, the formula defining "$x < y$" becomes $\exists z(y = x + s(z))$, the definition given above of "y is a prime" becomes the more perspicuous $\exists w(y = s(s(w))) \wedge \forall u \forall v(y = u \times v \supset (u = \underline{1} \wedge v = y) \vee (u = y \wedge v = \underline{1}))$, and the formalization of "there are infinitely many primes" becomes $\forall x \exists y(x < y \wedge y$ is a prime$)$.

A.2 The First Incompleteness Theorem

We are now in a position to formally define a "certain amount of arithmetic" sufficient for the first incompleteness theorem to apply to a formal system. In the following, the rules of reasoning of the formal systems we deal with will not be specified, but instead we will simply assume that any informal logical reasoning involving the language of arithmetic that we use can also be carried out within the system. This informal reasoning is in fact of a kind that can be formalized using the rules of first-order logic, so by the completeness theorem for first-order logic (see Chapter 7), we are justified in this assumption as long as the system encompasses those rules. We will also assume that the system satisfies the basic condition that its set of theorems is computably enumerable. (See Section 3.4.)

Suppose the language of such a theory includes the language of elementary arithmetic and that the axioms include the first six axioms of PA:

1. $\forall x \neg s(x) = 0$
2. $\forall x \forall y(s(x) = s(y) \supset = y)$
3. $\forall x(x + 0 = x)$

4. $\forall x \forall y (x + s(y) = s(x + y))$
5. $\forall x (x \times 0 = 0)$
6. $\forall x \forall y (x \times s(y) = x \times y + x)$

Consider a statement of the form "the Diophantine equation $D(x_1, \ldots, x_n) = 0$ has at least one solution (in natural numbers)." We assume this statement expressed in the language of PA, so that for example the equation "$x \times x - y \times y = 1$" is written "$x \times x = y \times y + 1$." Axioms (3), (4), (5), and (6) are all that is needed to prove every true statement of this form, since if an equation $s(x_1, \ldots, x_n) = t(x_1, \ldots, x_n)$ has a solution $k_1, \ldots, k_n$, axioms (3)–(6) prove $s(\underline{k}_1, \ldots, \underline{k}_n) = \underline{m}$ and $t(\underline{k}_1, \ldots, \underline{k}_n) = \underline{m}$ for some m, whereupon a simple logical inference leads to the conclusion that the equation has a solution. If T is consistent, it follows from the MRDP theorem that there are true statements of the form "the Diophantine equation $D(x_1, \ldots, x_n) = 0$ has no solution" that are not provable in T, as shown in Section 3.4. If, furthermore, T does not prove any *false* statement of the form "the Diophantine equation $D(x_1, \ldots, x_n) = 0$ has a solution," it follows that there are such statements that are undecidable in T.

Before commenting further on this result, let us observe that it is not really necessary for the language of T to *include* the language of arithmetic and the axioms of T to include (1)–(6). It suffices that s, $\times$ and $+$ can be *defined* in the language, and (1)–(6) proved restricted to the objects satisfying some formula $N(x)$. For example, in set theory we can define "x is a natural number" by a purely set-theoretical formula $N(x)$, that is, one that only uses the predicate "is a member of," and we can similarly define "the natural number x is the sum of the natural numbers y and z," "the natural number x is the successor of the natural number y," and "the natural number x is the product of the natural numbers y and z" and prove (1)–(6) using these definitions (together with a definition of 0). The reasoning by which incompleteness follows still goes through, with obvious modifications.

We can note further that the axioms (1) and (2) were not used in the argument. We need those axioms in order to arrive at Gödel's formulation of the incompleteness theorem. Axioms (1) and (2) suffice to *disprove* every false statement of the form "$\underline{m} = \underline{n}$," so together with (3)–(6) they disprove every false statement of the form "$D(\underline{k}_1, \ldots, \underline{k}_n) = 0$." Thus, given these axioms we can formulate the incompleteness theorem using Gödel's concept of ω-consistency: if it is not the case that there is an arithmetical formula

$A(x_1, \ldots, x_n)$ such that T proves "for some $x_1, \ldots, x_n$, $A(x_1, \ldots, x_n)$" but disproves every instance $A(\underline{k}_1, \ldots, \underline{k}_n)$, then T is incomplete. (Actually we only need the weaker assumption of Σ-soundness, formally defined below, since we only need to apply this condition in the special case of formulas of the form $D(x_1, \ldots, x_n) = 0$.)

Rosser's strengthening of the incompleteness theorem states that any *consistent* theory incorporating some basic arithmetic is incomplete. Replacing the assumption of ω-consistency or Σ-soundness by the assumption of consistency may seem a very minor strengthening of the incompleteness theorem. And indeed, if we consider those theories, like PA and ZFC, in which we formalize part of our mathematical knowledge, Σ-soundness is no more problematic than consistency, so the stronger form of the incompleteness theorem doesn't give us any new information about these theories. In theory, a skeptic might doubt the Σ-soundness of theories like PA or ZFC while being fairly convinced of their consistency, but in actuality such skeptics appear to be unknown.

Rosser's stronger version of the incompleteness theorem does, however, have significant applications, and in particular it patches up a weakness in the application of the theorem to theories in general, as presented in this book. In applying the incompleteness theorem to a theory T we isolate an "arithmetical component of T" within which we can define "x is a natural number," "$x \times y$," and so on, so as to be able to prove (1)–(6) in T restricted to the natural numbers. But it may well be that what we choose to regard as the arithmetical component of T is better interpreted in a different way, and is so interpreted by somebody who asserts the axioms of T. For this reason there need not be anything at all wrong with T even if it proves a statement of the form "the Diophantine equation $D(x_1, \ldots, x_n) = 0$ has a solution" that is false considered as a statement about the natural numbers. For example, suppose we have a theory T that proves not only (1)–(6), but also

$$\exists x(s(0) + x = x)$$

Regarded as an arithmetical statement, this sentence is trivially false: the equation $1 + x = x$ has no solution in natural numbers. But other interpretations of what we have chosen to regard as the arithmetical component of T may be perfectly reasonable. For example, a proponent of the theory T may interpret the variables as ranging, not over the natural numbers, but over the countable ordinals, which are a set-theoretical generalization of the natural numbers. On this interpretation, (1)–(6) are still true, but the equation $s(0) + x = x$ does have a solution, and in fact infinitely many solutions, among the infinite ordinal numbers.

Thus, using Rosser's strengthening of the incompleteness theorem in applying the theorem to a theory T, we can observe that although it may be more or less reasonable (depending on what else the theory proves) to describe a part of T within which "a certain amount of arithmetic" can be carried out as the "arithmetical component" of the theory, the theory will be incomplete with regard to statements within this component, unless it is inconsistent, which is a "bad" property however we interpret T.

Rosser's proof using a Rosser sentence for a theory T was briefly commented on in Section 2.7. There are several variants and generalizations of this proof, but none that applies the MRDP theorem in the straightforward way that yields Gödel's version of the incompleteness theorem. The Gödel-Rosser theorem is usually proved for theories that incorporate (in the sense indicated above) a slight strengthening of axioms (1)–(6), known as *Robinson arithmetic* (named after R. M. Robinson). One version of this theory uses an extension of the language of elementary arithmetic in which we do not define $<$, but instead include this symbol in the language from the outset, interpreting it as usual to denote the relation "is strictly smaller than." We extend axioms (1)–(6) with three axioms of which the first two do the same for the relation $<$ that axioms (1)–(6) do for addition, multiplication, and the successor function:

7. $\forall x \neg x < 0$
8. $\forall x \forall y (x < s(y) \equiv x < y \vee x = y)$
9. $\forall x \forall y (x < y \vee y < x \vee x = y)$

The interested reader is referred to Smullyan's book for a full and readable presentation of a proof, generalizing Rosser's argument, that any consistent theory incorporating axioms (1)–(9) is incomplete. Here we will consider Robinson arithmetic from the point of view of computability theory, in a formal definition of the concept of a Goldbach-like statement.

A.3 Goldbach-Like Statements, Σ-Formulas and Computably Enumerable Relations

Using $<$ we can define a special class of formulas containing only *bounded* quantification, which we will call *bounded* formulas. In such a formula, every universal quantification has the form $\forall x(x < t \supset A)$ and every existential quantification has the form $\exists x(x < t \wedge A)$ (where in both cases t is a term that does not contain x). We abbreviate these formulas as $\forall x < tA(x)$ and $\exists x < tA(x)$.

The essential property of bounded formulas is that there is an *algorithm* for deciding whether a sentence in which all quantifiers are bounded is true. To check whether a sentence $\forall x < tA(x)$ is true we compute the value n of t, and check whether $A(\underline{k})$ is true for every k smaller than n, and similarly for $\exists x < tA(x)$. Consider, for example, the sentence

$$\forall x < \underline{1000}(\underline{2} < x \wedge \exists w < x(x = \underline{2} \times w) \supset \exists y < x \exists z$$
$$< x(x = y + z \wedge y \text{ is a prime } \wedge z \text{ is a prime})$$

where we now define "y is a prime" by a formula containing only bounded quantifiers: $\underline{1} < y \wedge \forall u < y \forall v < y \neg u \times v = y$. By repeatedly applying the above method, we reduce the question of the truth or falsity of this sentence, which says that Goldbach's conjecture holds for every number smaller than 1000, to checking the truth or falsity of a finite number of sentences of the form $s = t$ or $s < t$. Similarly with any other sentence containing only bounded quantifiers. In fact, the reasoning by which we decide in this way on the truth or falsity of a bounded sentence can be carried out wholly within Robinson arithmetic.

It follows that any set E of natural numbers that can be defined by a bounded formula $A(x)$ is a computable set. For to check whether a number n is in E, we need only check whether the sentence $A(\underline{n})$ is true. The converse of this observation does not hold, that is, there are computable sets that cannot be defined by any bounded formula. But we can nevertheless formally characterize the computable sets using this concept, as will be seen in the following.

Note that the negation of a bounded formula can be formulated as a bounded formula, since $\neg\forall x < tA(x)$ is logically equivalent to $\exists x < t\neg A(x)$, and $\neg\exists x < tA(x)$ is logically equivalent to $\forall x < t\neg A(x)$. By repeated application of these rules, together with other logical equivalences such as the equivalence of $\neg(A \vee B)$ and $\neg A \wedge \neg B$, the negation of a bounded formula can be reduced to another bounded formula.

We can now give a formal definition of "Goldbach-like sentence" for the language of arithmetic. Quoting from Section 2.1:

> As we have seen, Goldbach's conjecture can be formulated as a statement of the form "Every natural number has the property

> P," where P is a computable property. This is a logically very significant feature of Goldbach's conjecture, and in the following, any statement of this form will be called a *Goldbach-like* statement. ... Actually, this description glosses over an important point: the property P must not only be computable, but must have a sufficiently simple form so that an algorithm for checking whether a number has the property P can be "read off" from the formulation of P.

We can now define a Goldbach-like sentence as a sentence of the form $\forall x A(x)$, where $A(x)$ is a bounded formula. As was noted above, this definition satisfies the condition that an algorithm for checking whether $A(\underline{n})$ is true for a particular number n can be formulated on the basis of the formulation of the condition $A(x)$.

While it is easy to see that Goldbach's conjecture itself can be formalized as a Goldbach-like sentence in this sense (just remove the restriction on $\forall x$ in the formula given), it is by no means obvious that every statement conforming to the informal definition of Goldbach-like given in Section 1.2 can be expressed as a Goldbach-like statement according to the given definition.

To see why this is so, we first introduce some further terminology. A formula of the form $\exists x A(x)$ where $A(x)$ contains only bounded quantification is called a Σ-*formula.* (A theory is Σ-*sound* if every Σ-sentence provable in the theory is true.) Suppose $\exists x A(x)$ has one free variable y, so that we write it $\exists x A(x, y)$. The set of n such that $\exists x A(x, \underline{n})$ is true is a computably enumerable set, since it can be enumerated by going through all sentences $A(\underline{m}, \underline{n})$, deciding their truth or falsity, and delivering n as output whenever $A(\underline{m}, \underline{n})$ is true. The converse of this also holds: for every computably enumerable set E there is a Σ-formula $\exists x A(x, y)$ such that E is the set of n for which $\exists x A(x, \underline{n})$ is true. To prove this requires a formalization of the informal definition of computably enumerable sets given in Chapter 3, for example in terms of Turing machines, together with an arithmetization of the formal definition. (The proof thus involves a great deal of detailed formal work, and so will not be attempted here.) In fact, if we are given a set E through some definition by which we know E to be computably enumerable (for example, "let E be the set of proofs in ZFC of the conjunction of some formula with its negation"), we can construct such a formula $\exists x A(x, y)$ which defines E. The statement "for every n, n is not in E" can then be formulated $\forall y \neg \exists x A(x, y)$, which can be equivalently expressed as $\forall z \forall y < z \forall x < z \neg A(x, y)$, and thus as a Goldbach-like statement according to the formal definition. In particular, given a property P that we know to be computably decidable, the statement "every natural number has property P" can be formulated as $\forall x A(x)$ for some bounded formula $A(x)$, since the complement of a decidable set is effectively computable, and the statement is equivalent to "for every n, n is not in the complement of the set of numbers having property P."

By the same kind of reasoning, applying the basic observation that a set E of numbers is computably decidable if and only if both E and its complement are computably enumerable, we find that the computably decidable sets are those which can be defined *both* by a Σ-formula $\exists x A(x, y)$ and by a Π-*formula*, a formula $\forall x A(x, y)$ with A bounded.

References

There is a large literature on the incompleteness theorem and on Kurt Gödel. The references given only cover writings that happen to have been referred to in the book. The chief reference for Gödel's scientific work is the *Collected Works*, while Dawson's book is the standard biography. Among electronic resources, the web pages of the Kurt Gödel Society (chartered in Vienna) contain many links relating to Gödel's life and work. The archives of the FOM (Foundations of Mathematics) mailing list, currently based at New York University, contain a wealth of discussions by logicians and philosophers relating to the incompleteness theorem, including the work of Harvey Friedman.

[Boolos 89] Boolos, G., "A New Proof of the Gödel Incompleteness Theorem," *Notices of the AMS* 36 (April 1989), pp. 388–390.

[Calude et al. 01] Calude, C. S., Hertling, P., Khoussainov, B., and Wang, Y., "Recursively Enumerable Reals and Chaitin Omega Numbers," *Theoretical Computer Science* 255 (2001), pp. 125–149.

[Chaitin 82] Chaitin, G. J., "Gödel's Theorem and Information," *International Journal of Theoretical Physics* 22 (1982), pp. 941–954.

[Chaitin 92] Chaitin, G. J., *Information-Theoretic Incompleteness*," World Scientific, Singapore, 1992.

[Chaitin 99] Chaitin, G. J., *The Unknowable*, Springer-Verlag, Singapore, 1999.

[Chaitin 05] Chaitin, G. J., "Irreducible Complexity in Pure Mathematics," electronic draft, March 24, 2005.

[Chaitin 05] Chaitin, G. J., *Metamath! The Quest for Omega.* To be published by Pantheon Books, New York, 2005.

[Chase and Jongsma 83] Chase, G. and Jongsma, C., *Bibliography of Christianity and Mathematics*, Dordt College Press, Sioux Center, IA, 1983.

[Dawson 97] Dawson, J., *Logical Dilemmas: The Life and Work of Kurt Gödel*, A K Peters, Natick, MA, 1997.

[Feferman 60] Feferman, S., "Arithmetization of Metamathematics in a General Setting," *Fundamenta Mathematica* 49 (1960), pp. 35–92.

[Feferman et al. 86] Feferman, S., Dawson, J. W., Goldfarb, W., Parsons, C., and Sieg, W. (eds.), *Kurt Gödel: Collected Works*, Vol. I–III, Oxford University Press, Oxford, 1986–1995.

[Feferman et al. 00] Feferman, S., Friedman, H., Maddy, P., and Steel, J., "Does Mathematics Need New Axioms?", *Bulletin of Symbolic Logic* 6 (2000), pp. 401–413.

[Franzén 04] Franzén, T., *Inexhaustibility: A Non-Exhaustive Treatment*, Lecture Notes in Logic 16, Association for Symbolic Logic and A K Peters, Wellesley, MA, 2004.

[Hofstadter 79] Hofstadter, D., *Gödel, Escher, Bach: An Eternal Golden Braid*, Basic Books, New York, 1979.

[Kadvany 89] Kadvany, J., "Reflections on the Legacy of Kurt Gödel: Mathematics, Skepticism, Postmodernism," *The Philosophical Forum* 20 (1989), pp. 161–181.

[Kline 80] Kline, M., *Mathematics: The Loss of Certainty*, Oxford University Press, Oxford, 1980.

[Kreisel 80] Kreisel, G., "Kurt Gödel," *Biographical Memoirs of Fellows of the Royal Society* 26 (1980), pp. 148–224.

[van Lambalgen 89] van Lambalgen, M., "Algorithmic Information Theory," *Journal of Symbolic Logic* 54 (1989), pp. 1389–1400.

[Li and Vitanyi 97] Li, M. and Vitanyi, P., *An Introduction to Kolmogorov Complexity and its Applications*, Springer-Verlag, New York, 1997.

[Lindström 01] Lindström, P., "Penrose's New Argument," *Journal of Philosophical Logic* 30 (2001), pp. 241–250.

[Lucas 61] Lucas, J. R., “Minds, Machines and Gödel,” *Philosophy* XXXVI (1961), pp. 112–127.

[Lucas 96] Lucas, J. R., “Minds, Machines, and Gödel: A Retrospect,” in P. J. R. Millican and A. Clark (eds.) *Machines and Thought: The Legacy of Alan Turing*, Vol. 1, Oxford University Press, Oxoford, 1996, pp. 103–124.

[Nagel and Newman 59] Nagel, E. and Newman, J., *Gödel's Proof*, Routledge & Kegan Paul, New York, 1959.

[Penrose 89] Penrose, R., *The Emperor's New Mind*, Oxford University Press, Oxford, 1989.

[Penrose 94] Penrose, R., *Shadows of the Mind*, Oxford University Press, Oxford, 1994.

[Penrose 96] Penrose, R., “Beyond the Doubting of a Shadow,” *PSYCHE* 2(23) (1996), electronic journal.

[Putnam 00] Putnam, H., “Nonstandard Models and Kripke's Proof of the Gödel Theorem,” *Notre Dame J. of Formal Logic* 41(1) (2000), pp. 53–58.

[Raatikainen 98] Raatikainen, P., “Interpreting Chaitin's Incompleteness Theorem,” *Journal of Philosophical Logic* 27 (1998), pp. 269–286.

[Rucker 95] Rucker, R., *Infinity and the Mind*, Princeton University Press, Princeton, NJ, 1995.

[Shapiro 03] Shapiro, S., “Mechanism, Truth, and Penrose's New Argument” *Journal of Philosophical Logic* 32 (2003), pp. 19–42.

[Shelah 03] Shelah, S., “Logical Dreams,” *Bulletin of the AMS* 40(2) (2003), pp. 203–228.

[Smullyan 92] Smullyan, R., *Gödel's Incompleteness Theorems*, Oxford University Press, Oxford, 1992.

[Sokal and Bricmont 98] Sokal, A. D. and Bricmont, J., *Fashionable Nonsense: Postmodern Intellectuals' Abuse of Science*, Picador Books, New York, 1998.

[Wang 87] Wang, H., *Reflections on Kurt Gödel*, MIT Press, Cambridge, MA, 1987.

Index